# REVISE AQA GCSE
# Geography A
## For the linear specification first teaching 2012

# REVISION WORKBOOK

Series Consultant: Harry Smith          Author: Rob Bircher

---

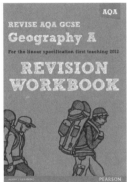

## THE REVISE AQA SERIES
**Available in print or online**

Online editions for all titles in the Revise AQA series are available Summer 2013.

Presented on our ActiveLearn platform, you can view the full book and customise it by adding notes, comments and weblinks.

### Print editions

Geography A Revision Workbook          9781447940890

Geography A Revision Guide          9781447940852

### Online editions

Geography A Revision Workbook          9781447940906

Geography A Revision Guide          9781447940869

For the linear specification first teaching 2012. This Revision Workbook is designed to complement your classroom and home learning, and to help prepare you for the exam. It does not include all the content and skills needed for the complete course.

To find out more visit www.[...]gcsegeographyrevision

ALWAYS LEARNING                    **PEARSON**

# Contents

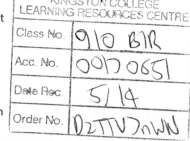
## A small bit of small print

AQA publishes Sample Assessment Material and the Specification on its website. That is the official content and this book should be used in conjunction with it. The questions in this book have been written to help you practise every topic in the book. Remember: the real exam questions may not look like this.

# Unstable crust

Study this map, which shows the Earth's tectonic plates.

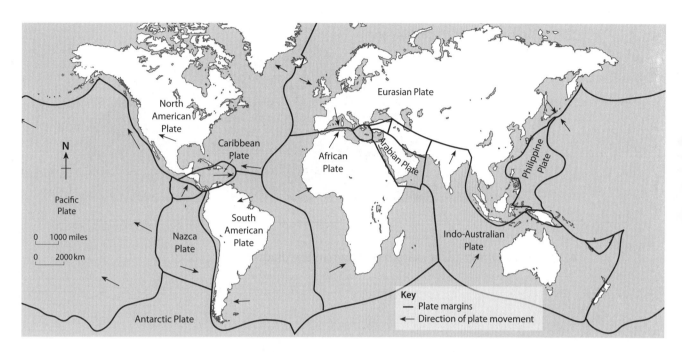

Guided

**1** Are the following statements about tectonic plates true or false?

**Tick the correct boxes.**

|  | True | False |
|---|---|---|
| Tectonic plates sit on top of the Earth's core |  | ✓ |
| Continental crust is lighter than oceanic crust |  |  |
| Oceanic crust is thicker than continental crust |  |  |
| Tectonic plates move due to convection currents in the mantle |  |  |

*(4 marks)*

Don't wory if maps and diagrams look different from ones you have revised; study them carefully and focus on what the question is asking you about them.

1

# Plate margins

Study this diagram, which shows a type of plate margin.

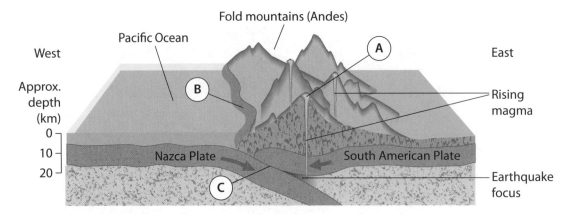

1  **(a)** What type of plate margin is shown in the diagram?

.......................................................................................................................................................

.......................................................................................................................................................

*(1 mark)*

**(b)** Some labels are missing from the diagram. Select the correct words from the following to identify points **A**, **B** and **C**.

| Shield volcano | Volcano | Oceanic trench |
|---|---|---|
| Mantle | | Subduction zone |

**Point A** ..................................................................................................................................................

**Point B** ..................................................................................................................................................

**Point C** ..................................................................................................................................................

*(3 marks)*

> This sort of question often has more terms than you need to answer the question. The extra terms will often be **nearly** right, so make sure you read them all carefully.

# Fold mountains and ocean trenches

Study this map, which shows the location of young fold mountains and ocean trenches.

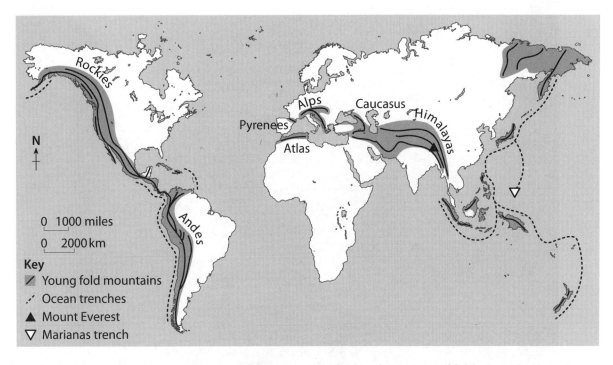

**1 (a)** What type of plate margin is usually responsible for the formation of fold mountains and ocean trenches?

.................................................................................................................................................

.................................................................................................................................................

*(1 mark)*

**(b)** Describe the stages in the formation of fold mountains.

The first stage is when layers of sediment are laid down in huge depressions called geosynclines and ......................................................................................................

.................................................................................................................................................

.................................................................................................................................................

.................................................................................................................................................

.................................................................................................................................................

.................................................................................................................................................

.................................................................................................................................................

.................................................................................................................................................

*(4 marks)*

# Fold mountains

This photograph shows a scene from a valley in a fold mountain range.

**1**   Use a named example to explain how people have adapted to the conditions found in a fold mountain range.

**Guided**

**Named example:** ............................................................................................................

One problem comes from the narrow valleys and the small amount of flat land for farming. The traditional solution to this in this region is to ..............................................

........................................................................................................................................

........................................................................................................................................

........................................................................................................................................

........................................................................................................................................

........................................................................................................................................

........................................................................................................................................

........................................................................................................................................

........................................................................................................................................

*(4 marks)*

Make sure that you read the question carefully. Here it asks for **conditions**, so you must cover more than one aspect of fold mountains.

# Volcanoes

Study this map which shows the location of the world's active volcanoes.

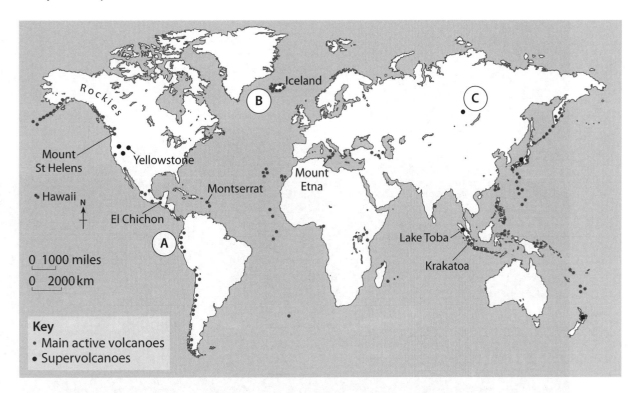

1 (a) Which of these statements best accounts for the location of active volcanoes at **A**, **B** and **C**?

Complete the table by writing the correct letter into the box beside each statement.

| Statement | Letter |
|---|---|
| These volcanoes are formed at a destructive plate margin | |
| These volcanoes are formed at a constructive plate margin | |
| This volcano is a long way from plate margins and is caused by something else (a rift valley) | |

*(2 marks)*

**Guided**

(b) Describe two differences between supervolcanoes and 'normal' volcanoes.

Supervolcanoes are much bigger than normal volcanoes with an eruption of at least 1000 km³ of magma ..............................................................................................................

..............................................................................................................................................

..............................................................................................................................................

..............................................................................................................................................

..............................................................................................................................................

..............................................................................................................................................

..............................................................................................................................................

*(4 marks)*

Remember that no credit is given for exact opposites.

5

  **Case study**

# Volcanoes as hazards

This photograph shows an eruption of Etna, a volcano in Sicily, Italy.

  **tier H**

**Guided**

**1** Explain why some volcanoes cause more damage and loss of life than others.

There are differences between volcanoes which make some eruptions more violent and destructive than others, and there are differences in where volcanoes are located which means their impact is bigger than in other areas. The reason why some volcanoes are more violent and destructive is ...............................................................

.................................................................................................................................

.................................................................................................................................

.................................................................................................................................

.................................................................................................................................

.................................................................................................................................

*(4 marks)*

'Explain' questions need you to give reasons.

# Earthquakes

Study this table of data about earthquakes in 2010.

| Location of earthquake | Magnitude on Richter scale | Number of deaths |
|---|---|---|
| Haiti | 7.0 | 316 000 (estimated) |
| Qinghai, China | 6.9 | 2698 |
| Maule region, Chile | 8.8 | 521 |
| Sumatra, Indonesia | 7.7 | 435 |
| Papua, Indonesia | 7.0 | 17 |

1  (a)  Which earthquake had the highest magnitude?

.................................................................................................................................. *(1 mark)*

(b)  Which earthquake had the highest death toll?

.................................................................................................................................. *(1 mark)*

(c)  Suggest **two** reasons why the earthquake with the highest magnitude did not cause the most deaths.

Reason 1 ......................................................................................................................

..................................................................................................................................

..................................................................................................................................

..................................................................................................................................

Reason 2 ......................................................................................................................

..................................................................................................................................

..................................................................................................................................

..................................................................................................................................

..................................................................................................................................

*(4 marks)*

> There are lots of different factors that control the effects of an earthquake. These can be divided into physical and human; for example, whether the earthquake has a shallow focus for physical, or whether it occurs in a densely populated region or not, for human.

# Earthquake hazards

This photograph shows damage done by an earthquake in Christchurch, New Zealand, in 2010.

**1** Explain why some earthquakes cause more damage and loss of life than others.

Earthquakes in poorer parts of the world often have a bigger impact than in richer parts of the world because buildings are not built to be earthquake-proof and because responses are not as quick and extensive as in richer parts of the world. Secondly, ...................................................................................................................

................................................................................................................................................

................................................................................................................................................

................................................................................................................................................

................................................................................................................................................

................................................................................................................................................

................................................................................................................................................

................................................................................................................................................

*(4 marks)*

**Case study**

# Tsunamis

**1** Use a named example to describe the effects a tsunami had on the coastal areas of a country or countries.

**2** Describe the effects a tsunami had on a coastal area of a country or countries.

> You only need to answer one of these questions.

⟩ **Guided** ⟩

**Named example:** ......................................................................................

The effects of the tsunami were very serious because the coastal areas were densely populated. The most serious impacts were the number of people killed and the effects on infrastructure, such as sewage systems being destroyed. The number of people killed was estimated at ......................................................................

..........................................................................................................................

..........................................................................................................................

..........................................................................................................................

..........................................................................................................................

..........................................................................................................................

..........................................................................................................................

..........................................................................................................................

*(6 marks (F), 8 marks (H))*

**Extra space**

..........................................................................................................................

..........................................................................................................................

..........................................................................................................................

..........................................................................................................................

..........................................................................................................................

..........................................................................................................................

..........................................................................................................................

..........................................................................................................................

> This question only asks you for '**effects**' so don't waste time giving 'causes' or 'responses'.

> When you use a case study in an answer, make sure you only include **relevant** details – don't just repeat everything you have learned about your case study.

# Rock groups

Study this map of rock types in the British Isles.

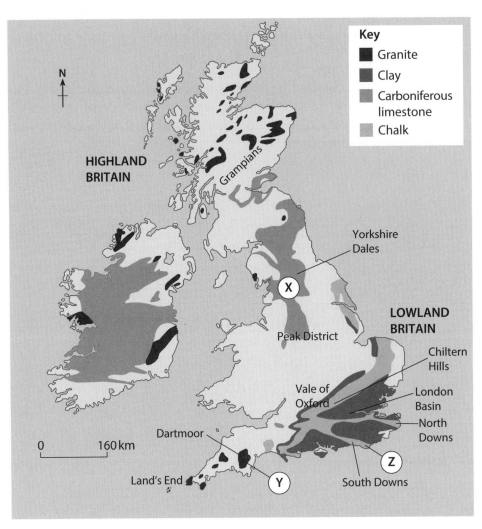

**Key**
- ■ Granite
- ■ Clay
- ■ Carboniferous limestone
- □ Chalk

1  **(a)** Complete the following table to describe the types of rock found at **X**, **Y** and **Z** on the map.

| Figure | Rock type |
|--------|-----------|
| X | Carboniferous limestone |
| Y | |
| Z | |

*(2 marks)*

**(b)** Using the map and your own knowledge, give an example of a sedimentary rock and an igneous rock.

**Sedimentary rock** ..............................................................................................................

**Igneous rock** .....................................................................................................................

*(2 marks)*

# The rock cycle

Study this photograph, which shows a cavern in Italy with stalactites and stalagmites.

**1 (a)** The formations shown in the photograph are associated with carbonation. What type of weathering is carbonation?

.................................................................................................................................................

*(1 mark)*

**(b)** Explain how carbonation works to weather limestone (calcium carbonate).

Carbonation changes calcium carbonate into calcium bicarbonate. First, carbon dioxide in the air combines with .....................................................................................

.................................................................................................................................................

.................................................................................................................................................

.................................................................................................................................................

.................................................................................................................................................

.................................................................................................................................................

.................................................................................................................................................

.................................................................................................................................................

*(4 marks)*

11

Had a go ☐   Nearly there ☐   Nailed it! ☐

**Case study**

# Granite landscapes

Study this photograph of Bowerman's Nose, a tor on Dartmoor.

**tier Fn**

**Guided**

**1** Describe the granite tor shown in the photograph and explain how it may have been formed.

The tor is a pillar of deeply jointed rock. It looks to be high up in the landscape, on the top of a hill. It is surrounded by scattered rocks. This tor probably formed when

........................................................................................................

........................................................................................................

........................................................................................................

........................................................................................................

........................................................................................................

........................................................................................................

........................................................................................................

........................................................................................................

........................................................................................................

........................................................................................................

........................................................................................................

*(6 marks)*

> There are two command words in this question – 'describe' and 'explain'. Make sure you do both in your answer.

# Carboniferous limestone landscapes

Study this diagram, which shows different features of a Carboniferous limestone landscape.

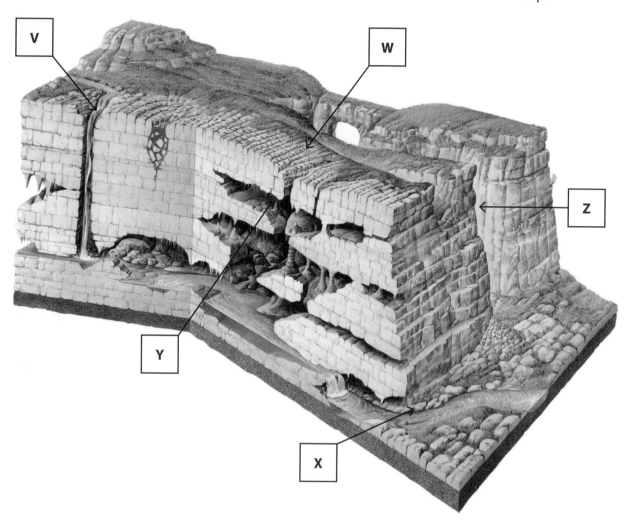

**1** The diagram has labels for different features at **V**, **W**, **X**, **Y** and **Z**. Complete this table by matching up each letter to the correct feature. Two have been completed for you.

| Carboniferous limestone feature | Letter |
|---|---|
| Cave | **Y** |
| Gorge | **Z** |
| Limestone pavement | |
| Resurgence stream | |
| Swallow hole | |

*(2 marks)*

> Don't rush questions like this: it is easy to make mistakes when you are filling in tables.

Case study

# Chalk and clay landscapes

This photograph shows a chalk escarpment and clay vale.

tier
F&H

**1** Label the photograph to show the location of:

  **(a)** the chalk escarpment (cuesta)

  **(b)** the clay vale

  **(c)** the scarp slope.

> Make sure the point of the arrow is touching the feature you are labelling.

*(3 marks)*

tier
F&H

**2** Complete this series of sketches illustrating the formation of a chalk escarpment (or cuesta) and a clay vale. The first stage has been completed for you.

Stage 1

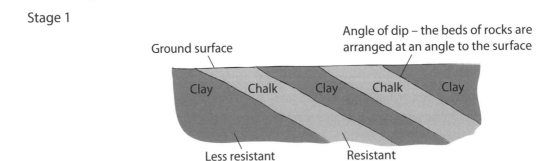

Angle of dip – the beds of rocks are arranged at an angle to the surface

Ground surface

Clay   Chalk   Clay   Chalk   Clay

Less resistant        Resistant

Stage 2

*(4 marks)*

# Quarrying

**Case study**

Study this map extract of Castleton in the Peak District. Hope Quarry is marked with an **X** on the map.

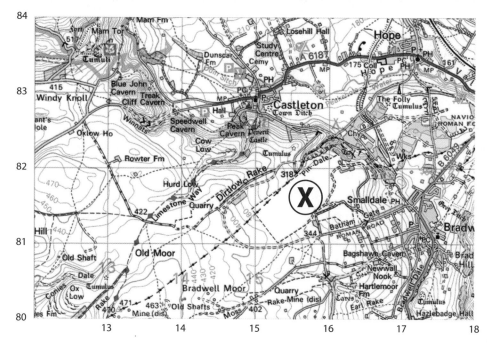

**Guided**

**1** Use the map to suggest environmental disadvantages of this quarry.

The quarry is very large; it takes up about one square kilometre on the map. This
means a large area of natural habitat has been destroyed and the site will make a big
impact on the local environment. Also ...........................................................................................

....................................................................................................................................................................

....................................................................................................................................................................

....................................................................................................................................................................

....................................................................................................................................................................

....................................................................................................................................................................

....................................................................................................................................................................

....................................................................................................................................................................

*(4 marks)*

**Extra space**

....................................................................................................................................................................

....................................................................................................................................................................

....................................................................................................................................

....................................................................................................................................

....................................................................................................................................

....................................................................................................................................

....................................................................................................................................

> Map extracts contain lots
> of detail you can use in
> a good answer. Provide
> accurate distances using
> the scale and refer to
> compass references
> for precise locations of
> features.

Had a go ☐     Nearly there ☐     Nailed it! ☐

# Quarrying management strategies

**Guided**

1 Using a named example of a quarry you have studied, describe how it has been managed to reduce negative impacts while it was in operation and how it has been restored once quarrying work was ended.

**Named example:** ...........................................................................................................................

While the quarry was in operation, planning controls restricted the expansion of the works to areas where the environmental impact was not going to be too great and

.............................................................................................................................................

.............................................................................................................................................

.............................................................................................................................................

.............................................................................................................................................

.............................................................................................................................................

.............................................................................................................................................

.............................................................................................................................................

.............................................................................................................................................

.............................................................................................................................................

.............................................................................................................................................

.............................................................................................................................................

*(8 marks)*

**Extra space**

.............................................................................................................................................

.............................................................................................................................................

.............................................................................................................................................

> When you use a case study in an answer, make sure you only include relevant details – don't just repeat everything you have learned about your case study.

16

# The UK climate

Study this map which shows average annual precipitation in the UK and Ireland.

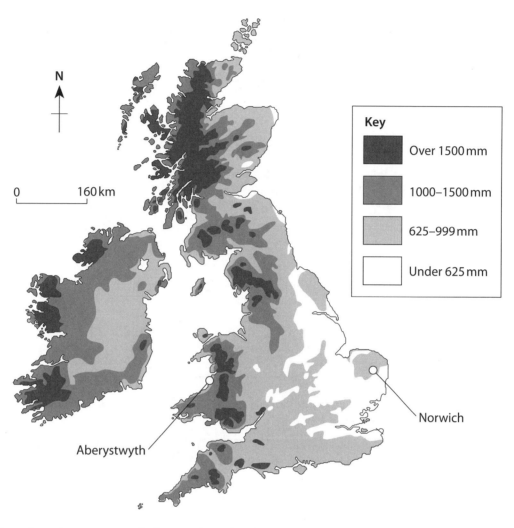

N

0 ⟶ 160 km

**Key**

Over 1500 mm

1000–1500 mm

625–999 mm

Under 625 mm

Norwich

Aberystwyth

**1 (a)** Complete the sentences below to describe the pattern of average annual precipitation in the UK.

**Add a compass direction in each space.**

Precipitation is highest in the ...north... and ...................... . As you move ...................., the amount of precipitation is reduced.

*(2 marks)*

**(b)** Suggest why the average annual precipitation for Aberystwyth is higher than for Norwich.

.................................................................................................................................................

.................................................................................................................................................

.................................................................................................................................................

.................................................................................................................................................

*(2 marks)*

# Depressions

Study this weather chart of western Europe.

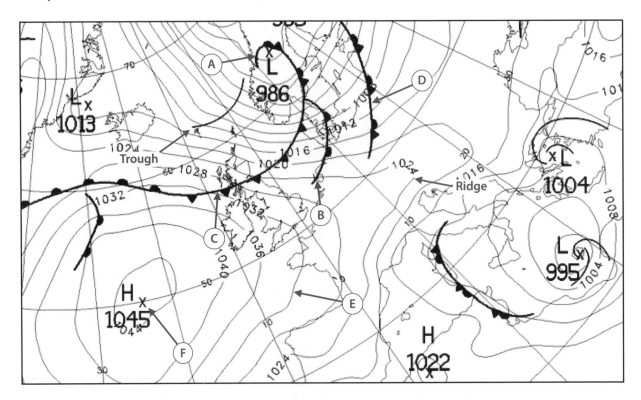

1  Complete the following table to identify each of the features **A**, **B**, **C**, **D**, **E** and **F** on this chart. Two have been completed for you.

| Feature | Letter |
| --- | --- |
| Isobar | E |
| Occluded front | D |
| Anticyclone | |
| Cold front | |
| Depression | |
| Warm front | |

*(4 marks)*

> Make sure you know the weather that is associated with the different fronts of a depression, and the weather conditions associated with anticyclones too.

# Anticyclones

Study this satellite photograph of the UK and western Europe during summer.

 tier F&H

**1 (a)** What weather system is shown dominating the majority of the UK in the photograph?

..................................................................................................................................................... *(1 mark)*

**(b)** Explain the summer weather conditions associated with this type of weather system.

..............................................................................................................................................................

..............................................................................................................................................................

..............................................................................................................................................................

..............................................................................................................................................................

..............................................................................................................................................................

..............................................................................................................................................................

..............................................................................................................................................................

*(4 marks)*

> The question asks about the **summer** weather conditions associated with the weather system shown in the photo. Read questions carefully so you can pick up this kind of important information.

**(c)** What is meant by the term 'blocking anticyclone'?

..............................................................................................................................................................

..............................................................................................................................................................

*(2 marks)*

**19**

# Extreme weather in the UK

Study this photograph of flooding in Tewkesbury, July 2007.

**tier H**

**Guided**

**1** Describe the different impacts that an extreme weather event can have.

Extreme weather events usually have negative impacts but they can have positive impacts too, such as communities coming together to help one another. Negative impacts include impacts on people's health, on ...................................................................

...........................................................................................................................................

...........................................................................................................................................

...........................................................................................................................................

...........................................................................................................................................

...........................................................................................................................................

...........................................................................................................................................

...........................................................................................................................................

*(4 marks)*

> Try to describe at least two impacts with development of both points for a good answer.

# Global warming

Study this graph of average global temperatures, 1860–2010.

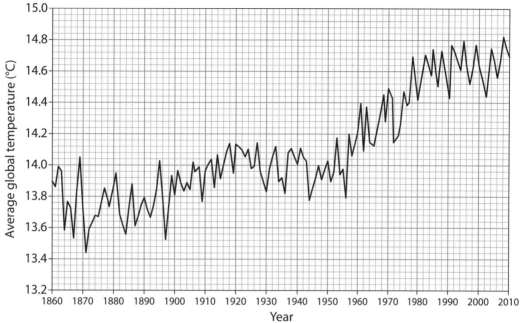

1  Which one of the following statements is correct? **Tick the correct box.**

☐ **A**    Average global temperature has fallen since 1860.

☐ **B**    Average global temperature was higher in 2010 than in 1860.

☐ **C**    Average global temperature has remained the same since 1860.

☐ **D**    Average global temperature was lower in 2010 than in 1860.          *(1 mark)*

2  Using the graph above, summarise the changes in average global temperature between
1910 and 2010.

..............................................................................................................................................

..............................................................................................................................................

..............................................................................................................................................

..............................................................................................................................................

*(2 marks)*

> Make sure you use numbers and
> data from the graph to develop
> your answer. Words such as 'rising'
> 'falling', 'trend', 'change', 'increase'
> and 'pattern' may also be useful
> to use in your answer. Look for
> any exceptions to the general
> trend of the graph.

# Consequences of global climate change

Study the following extract from a report on climate change in Bangladesh. It shows some of the predicted range of impacts from climate change on Bangladesh by 2100.

> Climate change can affect food production. Crop yields depend on temperature and rainfall and they may be reduced by up to 30 per cent. Cyclones will become stronger, with faster winds causing more damage; floods will become more common and along with a rise in sea level will mean many low lying areas will be flooded. This will lead to increased amounts of disease among the population.

**tier F**

**Guided**

1  (a)  List **two** of the predicted impacts of climate change on **people** in Bangladesh.

People may suffer from more disease. ..........................................................................................

...................................................................................................................................................

*(2 marks)*

**tier F&H**

**Guided**

(b)  Using the extract and your own knowledge, explain how climate change may affect the people of Bangladesh.

There are a range of impacts from climate change which will affect Bangladesh and its people. One of the most serious of these is the predicted drop in crop yields............

...................................................................................................................................................

...................................................................................................................................................

...................................................................................................................................................

...................................................................................................................................................

...................................................................................................................................................

...................................................................................................................................................

...................................................................................................................................................

...................................................................................................................................................

*(6 marks)*

**Extra space**

...................................................................................................................................................

...................................................................................................................................................

...................................................................................................................................................

...................................................................................................................................................

> Try to use information from the extract in your answer, then **develop** this information using your own knowledge and understanding.

# Responses to the climate change threat

Study the report from a local newspaper on climate change responses.

## Climate Change in Herefordshire

To reduce carbon emissions, Herefordshire Council will reduce waste and increase recycling and use less paper and water. It will help local homes and businesses become more energy efficient. It will reduce and control any pollution caused by council activities.

To prepare for possible consequences of climate change, the council will focus on improving urban drainage and upgrade flood management in case of increased flooding, invest in better water storage in case of drought, adapt forestry practices to take account of increased rainfall intensity, aim to compensate for biodiversity loss or changes and prepare for health risks caused by temperature increases.

**1** Use the newpaper report extract to describe **two** of the ways in which Herefordshire Council is responding to climate change.

1 ...........................................................................................................................................

2 ...........................................................................................................................................

*(2 marks)*

**2** Using the newspaper report, explain how climate change may affect Herefordshire.

> Guided

The council is preparing for a wide range of threats from climate change. Flooding is one high risk and the council says it is going to improve flood defences and also improve urban drainage and forestry practices. These are linked to the risk of flooding because ...............................................................................................

...........................................................................................................................................

...........................................................................................................................................

...........................................................................................................................................

...........................................................................................................................................

...........................................................................................................................................

...........................................................................................................................................

...........................................................................................................................................

*(4 marks)*

Try to develop the first point here and then you could make two more developed points, related to the information in the extract. This is where using your geographical understanding can help you make productive links between points.

# Causes of tropical revolving storms

Study this map, which shows the world distribution of tropical storms.

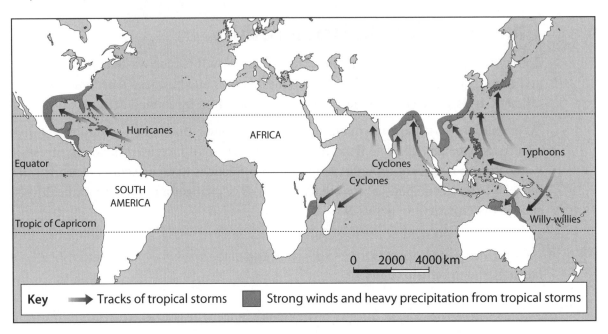

**Key**  → Tracks of tropical storms    ▮ Strong winds and heavy precipitation from tropical storms

Guided

**1** Using your own knowledge, explain the distribution of tropical revolving storms shown in the map.

The map shows the tropical revolving storms being largely restricted to the tropics. There are three main reasons for this location. First, these storms are powered by warm ocean temperatures – the sea water needs to be above 26.5 °C and these temperatures are only found in late summer and autumn in the tropics. Second, ..........

..................................................................................................................................................

..................................................................................................................................................

..................................................................................................................................................

..................................................................................................................................................

..................................................................................................................................................

..................................................................................................................................................

..................................................................................................................................................

..................................................................................................................................................

*(6 marks)*

**Extra space**

........................................................................

........................................................................

........................................................................

........................................................................

> You need to use your own knowledge to **explain** the distribution you see, rather than spending a lot of time describing what the map shows. Use the guided answer to get you started.

# Comparing tropical revolving storms

1  Tropical revolving storms often have different effects in rich and poor parts of the world. Contrast the effects of a named tropical revolving storm in a rich and in a poor part of the world, using one out of the following three categories of effect:

- economic effects

- social effects

- environmental effects.

...................................................................................................................................................

...................................................................................................................................................

...................................................................................................................................................

...................................................................................................................................................

...................................................................................................................................................

...................................................................................................................................................

...................................................................................................................................................

...................................................................................................................................................

...................................................................................................................................................

...................................................................................................................................................

...................................................................................................................................................

...................................................................................................................................................

...................................................................................................................................................

...................................................................................................................................................

...................................................................................................................................................

...................................................................................................................................................

*(8 marks)*

**Extra space**

...................................................................................................................................................

...................................................................................................................................................

...................................................................................................................................................

...................................................................................................................................................

Try to link your accounts by using connective words such as 'whereas' or 'on the other hand'.

This question asks for a named example. This means a named example for both storms: the rich area one and the poor area one, so make sure you do this. Contrast means you need to describe the difference in the effects in the rich and poor world.

# What is an ecosystem?

Study this diagram of a pond ecosystem.

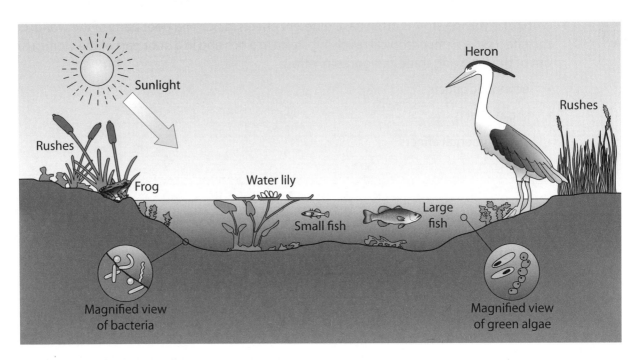

**1** What is meant by the term ecosystem?

 **Guided**

A community of living organisms (plants and animals) and their physical environment (sunlight, air, water, rock and soil) which are all linked together and often depend on each other. ................................................................................................................

*(2 marks)*

**2** Use the diagram above to identify one example for each of the following parts of the pond ecosystem:

**(a)** a producer: ..............................................................................................................

*(1 mark)*

**(b)** a consumer: .............................................................................................................

*(1 mark)*

**(c)** a decomposer: ..........................................................................................................

*(1 mark)*

> Try not to rush one mark questions like these – it is easy to make a silly mistake.

# Ecosystems

**1 (a)** Name the approximate latitude of the world's hot deserts.

...................................................................................................................................................

*(1 mark)*

**(b)** Explain how vegetation in a hot, dry desert ecosystem is adapted to climate and soil.

Hot, dry desert conditions pose significant challenges to plant life because of the severe lack of water (less than 250 mm of rain per year) and high daytime temperatures (often between 35 and 45 °C degrees in summer), which make transpiration rates very high. Desert soils are also very dry and because plant cover is sparse, they have very little organic content and can be salty as well. Plants adapt to these conditions in a range of ways. Some store water in their stems, for example cacti. This means ..........................................................................................

...................................................................................................................................................

...................................................................................................................................................

...................................................................................................................................................

...................................................................................................................................................

...................................................................................................................................................

...................................................................................................................................................

...................................................................................................................................................

*(6 marks)*

**Extra space**

...................................................................................................................................................

...................................................................................................................................................

...................................................................................................................................................

...................................................................................................................................................

> Your answer needs to explain how plants are adapted to the hot dry desert conditions, not just describe their features.

> Also remember that both **climate** and **soil** need to be covered.

 Case study

# Temperate deciduous woodlands

Study the extract about management in Epping Forest.

> Epping Forest is an ancient area of deciduous forest land in north-east London. It covers an area of around 2500 hectares. Epping Forest is a Site of Special Scientific Interest which also protects it from development. It is managed by the City of London Corporation which aims to both protect the forest and make it available to people to use for recreation. The corporation uses traditional techniques to encourage new tree growth and maintain the natural ecosystem. For example, pollarding is used, chopping off branches so that trees sprout more new branches. Dead wood is allowed to rot away and grassy areas are left uncut, which both encourage wildlife. Traditional meadows are cut. Some types of recreation are controlled to protect some areas from being damaged.

**1** Use the extract and your own knowledge to explain how Epping Forest is being managed in a sustainable way.

'Sustainable' means that the forest is being managed in a way that meets the needs of people now, yet will allow future generations to use it in the same way. So techniques like pollarding are sustainable because wood is being taken from trees in a way that actually encourages the trees to grow more and live longer. ........................

.................................................................................................................................

.................................................................................................................................

.................................................................................................................................

.................................................................................................................................

.................................................................................................................................

.................................................................................................................................

.................................................................................................................................

.................................................................................................................................

.................................................................................................................................

*(6 marks)*

**Extra space**

.................................................................................................

.................................................................................................

.................................................................................................

.................................................................................................

> Make sure you focus on 'how', not 'why'. Your answer should also link the features you describe with the things that make them sustainable. You can also bring in strategies that are not included in the extract, such as controlled felling and replanting.

# Deforestation

**1 (a)** Deforestation in a tropical rainforest may have positive and negative effects for people living in the area. Describe these effects.

Deforestation creates land for farming and for building roads and houses. These all benefit local people by giving them ways to earn a living and places to live. Deforestation may also create jobs for local people if it is associated with mining, logging, ranching, etc. But deforestation also comes at a cost. .......................................

..............................................................................................................................

..............................................................................................................................

..............................................................................................................................

..............................................................................................................................

..............................................................................................................................

..............................................................................................................................

*(4 marks)*

> The guided answer gives some benefits; to finish the answer off you should include a couple of negative impacts. The question asks for impacts on people living in the area, so it should not be things like increased carbon dioxide in the atmosphere as that has a global consequence, not a local one.

**(b)** The following is a list of some of the causes of deforestation.

- commercial farming and ranching
- mineral extraction
- logging

Choose **two** of the causes listed or others that you have studied. Explain how your choices cause deforestation.

Cause 1 ....................................................................................................

..............................................................................................................................

..............................................................................................................................

..............................................................................................................................

..............................................................................................................................

Cause 2 ....................................................................................................

..............................................................................................................................

..............................................................................................................................

..............................................................................................................................

..............................................................................................................................

*(4 marks)*

# Sustainable management of tropical rainforest

Read the following newspaper article.

> The South American country of Ecuador has rich rainforest resources, which are the most biodiverse in the world. But huge reserves of oil have also been discovered beneath the Yasuni tropical rainforest national park – worth $7.2 billion. Instead of extracting this oil, the Ecuador government has asked the world to pay it half that amount in order to 'leave the oil in the soil'. Environmentalists have praised the Ecuador government for this proposal; after all, it is the rich countries that have damaged the world most with pollution. But others have criticised the scheme, saying it is nothing more than environmental extortion.

**1 (a)** Study the following statements about the newspaper article.

Complete the table by ticking the correct box to show whether each statement is **True** or **False**.

| Statement | True | False |
|---|---|---|
| Extracting the oil from Ecuador's Yasuni National Park would damage the rainforest. | ✓ | |
| Ecuador has offered to save its rainforest in return for $7.2 billion from richer countries. | | |
| The Ecuador proposal is one example of how rainforest can be conserved for the future. | | |
| There is international agreement that the Ecuador proposal is a very good idea. | | |

*(4 marks)*

**(b)** Name **two** other ways tropical rainforest can be managed apart from conservation swaps such as that described in the newspaper article.

......................................................................................................................................................

......................................................................................................................................................

*(2 marks)*

> Questions like this will never try to trick you into making the wrong choice, but you need to check each statement carefully against the extract. There's no point rushing your answer and making a mistake.

Case study **Rainforest management**

Study this graph, which shows data on the rate of deforestation in the Brazilian Amazon.

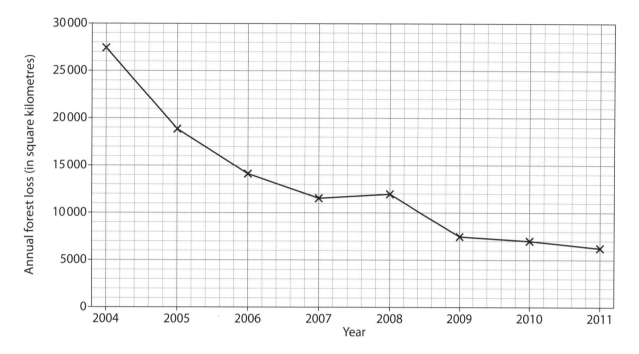

**1 (a)** Describe the trend shown by the deforestation data in the graph.

Guided

The rate of deforestation has declined almost every year between ...............................

..................................................................................................................

*(1 mark)*

**(b)** Using your own knowledge of sustainable management of tropical rainforests, suggest **three** reasons for the decline in deforestation rates shown in the graph.

Reason 1 ...........................................................................................................

..................................................................................................................

Reason 2 ...........................................................................................................

..................................................................................................................

Reason 3 ...........................................................................................................

..................................................................................................................

*(3 marks)*

# Case study Economic opportunities in hot deserts 1

1   Describe how people use a hot desert area in a richer part of the world to make a living.

..................................................................................................................................................................

..................................................................................................................................................................

..................................................................................................................................................................

..................................................................................................................................................................

..................................................................................................................................................................

..................................................................................................................................................................

..................................................................................................................................................................

..................................................................................................................................................................

..................................................................................................................................................................

..................................................................................................................................................................

..................................................................................................................................................................

..................................................................................................................................................................

..................................................................................................................................................................

..................................................................................................................................................................

..................................................................................................................................................................

*(6 marks)*

**Extra space**

..................................................................................................................................................................

..................................................................................................................................................................

..................................................................................................................................................................

..................................................................................................................................................................

> Make sure you focus on the economic opportunities of your case study, such as mining, agriculture and tourism, and not the challenges or management techniques you may have learned about, because this question does not ask for those.

# Economic opportunities in hot deserts 2

**1** Describe the challenges faced by people in hot desert areas and show how they are different between rich and poor parts of the world. Refer to suitable examples in your answer.

..................................................................................................................................

..................................................................................................................................

..................................................................................................................................

..................................................................................................................................

..................................................................................................................................

..................................................................................................................................

..................................................................................................................................

..................................................................................................................................

..................................................................................................................................

..................................................................................................................................

..................................................................................................................................

..................................................................................................................................

..................................................................................................................................

..................................................................................................................................

..................................................................................................................................

..................................................................................................................................

..................................................................................................................................

..................................................................................................................................

*(8 marks)*

**Extra space**

..................................................................................................................................

..................................................................................................................................

..................................................................................................................................

..................................................................................................................................

> Make sure you start with the general challenges faced by people living in hot desert areas and then explore the differences between richer and poorer areas. This question asks for examples so try to include names and some details.

# Changes in the river valley

Study this diagram, which shows the drainage basin of a river.

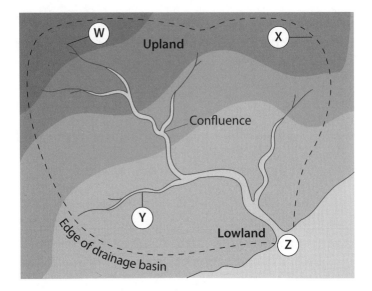

**1 (a)** Complete the following table by writing one of the letters **W**, **X**, **Y** or **Z** against the correct map label. One box will be left empty.

| Label | Letter |
|-------|--------|
| Tributary | |
| Mouth | |
| Source | |
| Meander | |
| Watershed | |

*(4 marks)*

**(b)** Complete the sentences below to describe the way a river's long profile and cross profile change over its course.

| ~~upland~~ | cross profile | long profile | steep |
|:---:|:---:|:---:|:---:|
| **flat** | **lowland** | **vertical** | **lateral** |

Rivers begin in ...upland... areas and flow downhill. Near the source the ............................................

of a river shows a steep gradient. It gradually gets lower and less steep until the river

reaches sea level. The river has a V-shaped .................................................................. in

the upper course. By the time the river reaches its lower course, the valley is wide and ...........

.

*(3 marks)*

The first gap has been filled – now work out which are the correct items to complete the rest of the passage.

# Erosion and transportation

Study this diagram, which shows the different ways a river transports its load.

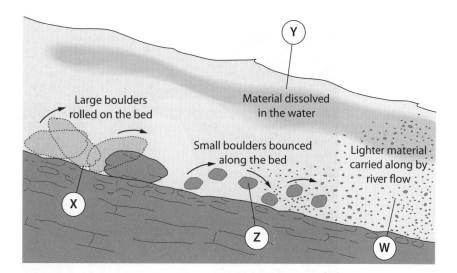

1   Complete the following table by writing **one** of the letters **W**, **X**, **Y** or **Z** against the correct diagram label. One box will be left empty.

| Label | Letter |
|---|---|
| abrasion | |
| saltation | |
| solution | |
| suspension | |
| traction | |

*(4 marks)*

2   **(a)** What is 'discharge'?

**Guided**

Discharge is the volume or flow of water passing a river measuring station at a
particular time. ...................................................................................................................................................

*(1 mark)*

**(b)** Describe how the rate of discharge affects river erosion and deposition.

.............................................................................................................................................................................

.............................................................................................................................................................................

.............................................................................................................................................................................

.............................................................................................................................................................................

.............................................................................................................................................................................

*(2 marks)*

# Waterfalls and gorges

Study this photograph of a waterfall in the Brecon Beacons National Park in Wales.

**1 (a)** Describe **three** features of the waterfall shown in the photograph.

.......................................................................................................................................

.......................................................................................................................................

.......................................................................................................................................

.......................................................................................................................................

.......................................................................................................................................

.......................................................................................................................................

*(3 marks)*

**Guided**

**(b)** Draw a labelled diagram to explain how a waterfall is formed.

> This answer has been started for you, but in the exam for a question like this, you would be drawing it from scratch.

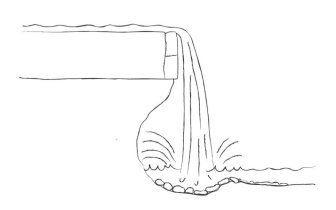

*(4 marks)*

**36**

# Erosion and deposition

Study this photograph of part of the River Severn in Shropshire.

meander neck

**Guided**

**1** Label the following features on the photograph.

> One label has been done for you.

~~meander neck~~         **river cliff**         **slip-off slope**         **floodplain**

*(4 marks)*

**2** Draw a labelled diagram(s) to explain how an ox-bow lake is formed.

> If a question says 'Draw a diagram', this doesn't mean you can only draw one thing – you could draw a picture for each stage in the formation of the ox-bow lake. Make sure your labels show the sequence of events.

*(4 marks)*

# Flooding 1

Study this diagram, which shows discharge after the same rainstorm in two drainage basins.

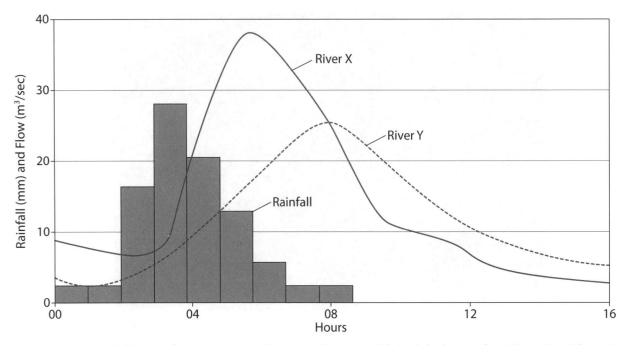

**1 (a)** Label the following features on the diagram. You can add the labels to either **River X** or **River Y**.

**rising limb**                    **falling limb**                    **lag time**

*(3 marks)*

**(b)** Which river, **River X** or **River Y**, is most likely to flood?

..................................................................................................................................................

*(1 mark)*

> **Guided**

**(c)** The drainage basin for **River X** is mostly moorland, while the drainage basin for **River Y** is mostly forested. Use this information to account for the differences between their two hydrographs.

The moorland in River X's drainage basin would intercept rainfall less than the forest land use in River Y's drainage basin. This means .................................................................

..................................................................................................................................................

..................................................................................................................................................

*(2 marks)*

> Remember that different factors affect discharge, including land use.

38

# Flooding 2

Study the table below, which shows a list of some major river flooding events in the UK.

| Date | Flood event and location | Most severe impact |
|------|--------------------------|--------------------|
| 1864 | Great Sheffield Flood, Sheffield | 270 people killed |
| 1947 | Great floods – many areas affected | 100 000 properties flooded |
| 1952 | Lynmouth flood, Devon | 34 people killed |
| 1979 | South Wales floods | 100s of homes and businesses flooded |
| 1998 | Easter floods, Midlands | 5 people killed |
| 2000 | Flooding across England | 10 000 homes and businesses flooded |
| 2002 | Glasgow floods | 200 people evacuated |
| 2004 | Boscastle flood, Cornwall | 100 people airlifted to safety |
| 2005 | Floods in Carlisle, Cumbria | 3 people killed |
| 2007 | UK floods – many areas affected | 6 people killed |
| 2009 | Cumbria floods | 1 person killed |
| 2010 | Flooding in Cornwall | 100 people evacuated |

1   Which of the following statements best describes the trend suggested by the table? Tick one box to show your answer.

| | Statement | |
|---|-----------|---|
| A | River flooding is getting less frequent in the UK than it used to be. | |
| B | River flooding happens more or less as often now as it did in previous decades. | |
| C | River flooding events in the UK appear to be more common than they used to be. | |

*(1 mark)*

> Guided

2   Suggest reasons why the responses to flood events often seem to be different in poorer areas of the world compared with richer areas.

One big difference in responses to flood events in poorer countries compared with richer countries is that in poorer countries international aid is often needed to help fund and deliver the response to a major flood. This aid takes time to arrange, delaying some aspects of the response. In richer parts of the world, the government can usually fund and organise a quick and effective response. This difference is due to ...............................................................................................................................................

.........................................................................................................................................................................

.........................................................................................................................................................................

.........................................................................................................................................................................

.........................................................................................................................................................................

.........................................................................................................................................................................

*(6 marks)*

# Hard and soft engineering

Read this news article about flood defences in Boscastle.

In August 2004 the River Valency flooded Boscastle, causing huge amounts of damage. Since 2004, flood walls have been built along parts of the riverbank. The river has been widened and deepened in places. The old bridge, which trapped debris and made flooding worse in 2004, has been replaced with a new higher and wider bridge. The village car park, which saw cars washed away during the flood, has been moved onto higher ground and hedges have been planted to divide the car park into sections. When floods came to Boscastle again in 2007, the impact was much less than in 2004.

1   Using the information in the news article above, complete the following table to show whether the flood management strategies are hard engineering solutions or soft engineering solutions. Use a tick to show your answer.

| Flood management strategies | Hard engineering | Soft engineering |
| --- | --- | --- |
| Building flood walls | ✓ | |
| Planting hedges to divide up the car park | | |
| Widening and deepening the river | | |
| Moving the car park to a higher location | | |

*(4 marks)*

2   There is often debate over whether to use hard engineering or soft engineering to manage flooding. Put forward a case for **either** hard engineering or soft engineering being the best choice.

Chosen strategy ...................................................................................................................................

.............................................................................................................................................................

.............................................................................................................................................................

.............................................................................................................................................................

.............................................................................................................................................................

.............................................................................................................................................................

.............................................................................................................................................................

.............................................................................................................................................................

.............................................................................................................................................................

.............................................................................................................................................................

.............................................................................................................................................................

.............................................................................................................................................................

.............................................................................................................................................................

.............................................................................................................................................................

*(8 marks)*

**Case study** # Managing water supply

Read this blog post from March 2012 about low water levels in Rutland Water reservoir.

> For two years now we have had really dry conditions in winter as well as spring and summer and visitors to Rutland Water, Britain's largest man-made reservoir, cannot fail to notice that water levels in the reservoir are getting very low. Since the reservoir supplies the drinking water to millions of people in the Midlands, I guess we should all think hard about how we could use less water (or just hope that 2012 has the wettest summer on record!). Anglian Water has applied to abstract more water from the River Nene, that is to take out water from the river and use it to fill up the reservoir, but what impacts would that have on the river, I wonder, and the wildlife that depends on it?

**1** Using the information in the blog post, and your own knowledge, suggest why water agencies have to get special permission to take lots of water out of rivers to fill up reservoirs during dry periods.

The discharge of a river has a big impact on erosion and deposition. Reducing the volume of water in a river would slow the river down and mean more of the river's load was deposited. This might raise the riverbed in places and lead to possible flooding problems in the future. Also, many organisms depend on the river, such as

..............................................................................................................................................

..............................................................................................................................................

*(4 marks)*

**2** Use a case study to describe the economic, social and environmental consequences of building a reservoir.

**Name of reservoir** ..........................................................................................................................

..............................................................................................................................................

..............................................................................................................................................

..............................................................................................................................................

..............................................................................................................................................

..............................................................................................................................................

..............................................................................................................................................

..............................................................................................................................................

..............................................................................................................................................

..............................................................................................................................................

..............................................................................................................................................

..............................................................................................................................................

..............................................................................................................................................

*(8 marks)*

# Changes in ice cover

Study this diagram, which shows carbon dioxide levels in the atmosphere during glacial and interglacial periods for the last 650 000 years.

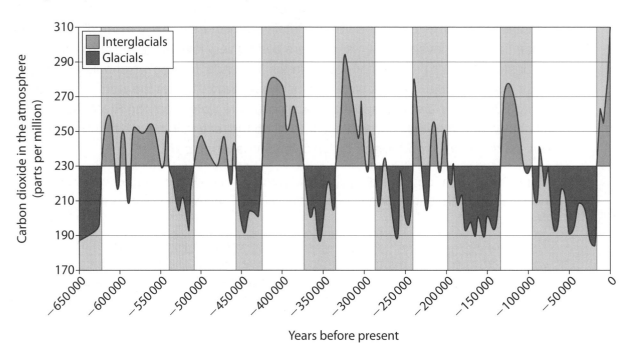

Years before present

**1** **(a)** Describe the link between atmospheric carbon dioxide and the occurrence of glacials and interglacials.

..................................................................................................................................................

..................................................................................................................................................

..................................................................................................................................................

*(1 mark)*

**(b)** Explain how ice core samples can provide evidence of atmospheric carbon dioxide levels in the past.

*Ice sheets are made up of layers of ice going back many thousands of years. Each new layer of ice that developed contained bubbles of atmospheric gases from that period. So* ...........................................................................

..................................................................................................................................................

..................................................................................................................................................

..................................................................................................................................................

..................................................................................................................................................

..................................................................................................................................................

..................................................................................................................................................

..................................................................................................................................................

*(4 marks)*

 **Case study**

# The glacial budget

Study the series of photographs taken of the Grinnell Glacier in Montana, USA, from the same spot at the same time of year in 1938, 1981, 1998 and 2009.

*Grinnell Glacier, 1938*     *Grinnell Glacier, 1981*     *Grinnell Glacier, 1998*     *Grinnell Glacier, 2009*

**1 (a)** As well as photographs, name **two other** types of evidence that geographers can use to show that glaciers have retreated.

Evidence type 1: ........................................................................................................................

........................................................................................................................

Evidence type 2: ........................................................................................................................

........................................................................................................................

*(2 marks)*

> Remember that evidence of glacial retreat can include landforms as well as data.

**(b)** What is meant by the term 'negative glacial budget'?

........................................................................................................................

........................................................................................................................

*(1 mark)*

> For 'what is meant by' questions you just need to provide a definition, not an explanation.

# Glacial weathering, erosion, transportation and deposition

Study this photograph of the Grosser Aletsch glacier in the Alps.

freeze–thaw
weathering

**1 (a)** Label the diagram to show where you would expect to find evidence of:

One label has been done for you.

    **(i)** freeze–thaw weathering

    **(ii)** transportation

    **(iii)** deposition.

*(3 marks)*

**(b)** Draw a diagram to explain how glaciers erode by **plucking**.

*(2 marks)*

**(c)** Describe the glacial transport process known as bulldozing.

........................................................................................................................................................

........................................................................................................................................................

*(2 marks)*

# Glacial erosion landforms 1

**1** Select the correct words to complete this paragraph about the formation of a corrie.

slip   erode   expands   weathering   arête   gap   ~~compacts~~   abrasion

Snow in a mountainside hollow ..compacts.. into ice. Freeze–thaw ................................ around

the corrie means rock falls onto the ice. These rocks help ............................ the base of the corrie.

A lip forms where the ice leaves the corrie because of rotational ..............:

*(3 marks)*

> The first term has been added for you. Work out which are the correct terms to complete the rest of the passage

**2** Name **two** features that would help identify a corrie on an OS map extract of a previously glaciated area.

.................................................................................................................................................

.................................................................................................................................................

.................................................................................................................................................

.................................................................................................................................................

*(2 marks)*

**3** Draw a labelled diagram to illustrate how corries form.

*(4 marks)*

# Glacial erosion landforms 2

Study this OS map extract of the Langdale Valley in the Lake District (1:50 000 scale enlarged).

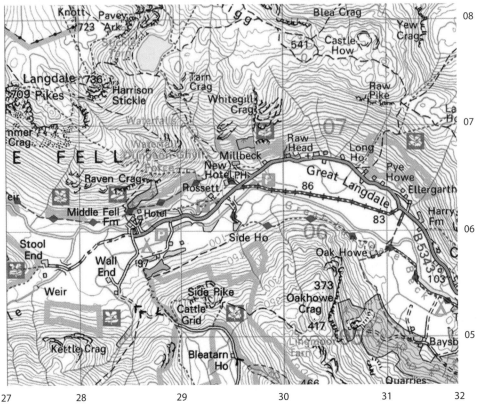

**1** Describe the evidence you can see in this map extract that suggests that the Langdale Valley has been glaciated.

..................................................................................................

..................................................................................................

.  ..................................................................................................

..................................................................................................

..................................................................................................

..................................................................................................

..................................................................................................

..................................................................................................

..................................................................................................

..................................................................................................

..................................................................................................

..................................................................................................

..................................................................................................

> Look for evidence of a glacial trough (steep sides, flat bottom), misfit streams (much too small for the valley they are in), truncated spurs, hanging valleys (often marked now by waterfalls). Use grid references and / or named locations on the map.

*(6 marks)*

46

# Glacial landforms of transportation and deposition

Study this photograph of the Glacier de Moiry, in Switzerland.

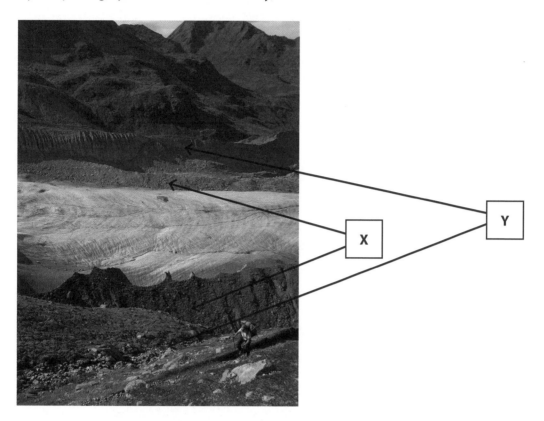

1  Identify the landform of glacial transportation and deposition labelled **X** in the photograph. Use a tick to show your choice.

*(1 mark)*

| Landform | |
|---|---|
| Terminal moraine | |
| Drumlin | |
| Erratic | |
| Lateral moraine | |

2  The landform labelled **Y** is a second, higher range of ridges either side of the glacier. Suggest how features X and Y were formed.

Landform Y probably formed when the glacier was wider than it is currently. Like X, landform Y is a .................................................................................................................................

.............................................................................................................................................................

.............................................................................................................................................................

.............................................................................................................................................................

.............................................................................................................................................................

.............................................................................................................................................................

# Alpine tourism – attractions and impacts

**1** Use a named Alpine area to describe the environmental impacts of tourism.

**Name of Alpine area:** ................................................................................................................................

...........................................................................................................................................................................

...........................................................................................................................................................................

...........................................................................................................................................................................

...........................................................................................................................................................................

...........................................................................................................................................................................

...........................................................................................................................................................................

...........................................................................................................................................................................

...........................................................................................................................................................................

...........................................................................................................................................................................

...........................................................................................................................................................................

...........................................................................................................................................................................

...........................................................................................................................................................................

...........................................................................................................................................................................

...........................................................................................................................................................................

...........................................................................................................................................................................

...........................................................................................................................................................................

*(8 marks)*

**Extra space**

...........................................................................................................................................................................

...........................................................................................................................................................................

...........................................................................................................................................................................

...........................................................................................................................................................................

...........................................................................................................................................................................

> To do really well on a question like this you need to use plenty of relevant detail from your case study rather than just making generic statements such as 'tourism causes damage to footpaths'.

# Management of tourism and the impact of glacial retreat

**1** Use a **case study** of an Alpine area to describe how people try to manage tourism in the area in a sustainable way.

.......................................................................................................................................

.......................................................................................................................................

.......................................................................................................................................

.......................................................................................................................................

.......................................................................................................................................

.......................................................................................................................................

.......................................................................................................................................

.......................................................................................................................................

.......................................................................................................................................

.......................................................................................................................................

.......................................................................................................................................

.......................................................................................................................................

.......................................................................................................................................

.......................................................................................................................................

*(6 marks (F), 8 marks (H))*

**Extra space**

.........................................................................................................

.........................................................................................................

.........................................................................................................

.........................................................................................................

.........................................................................................................

.........................................................................................................

.........................................................................................................

> Make sure you name your case study and try to link what you say about the issues caused by tourism in the area to the way these issues are managed. You also need to include sustainability – how tourism is managed so that future generations will be able to use the resources in the same way, and includes the idea of tourism benefiting local communities first and foremost.

# Waves and coastal erosion

Study this newspaper article.

> A woman has died after being buried by a cliff collapse in Dorset. Witnesses report that a small landslide was immediately followed by a much larger one which deposited around 400 tonnes of rock and mud from the top of the cliff onto the beach below. The accident happened on a stretch of Dorset's Jurassic coast which is composed of soft sandstone. A few days before the accident, Dorset Council had warned visitors to the coast to stay well away from the cliff at all times due to the high risk of landslides and mudflows due to many weeks of heavy rainfall in the area.

**1** **(a)** Which of the following geographical terms best describes the process which caused this fatal accident? Use a tick to indicate your choice.

*(1 mark)*

| Geographical term | |
|---|---|
| Weathering | |
| Attrition | |
| Mass movement | |
| Saltation | |

> Guided

**(b)** Using the newspaper article and your own knowledge, describe **two** factors which contributed to this landslide.

Factor 1: The wet weather will have saturated .................................................................................

............................................................................................................................................................

............................................................................................................................................................

Factor 2: ............................................................................................................................................

............................................................................................................................................................

............................................................................................................................................................

............................................................................................................................................................

*(4 marks)*

# Coastal transportation and deposition

**1 (a)** Draw a labelled diagram to explain the process of longshore drift.

> Guided

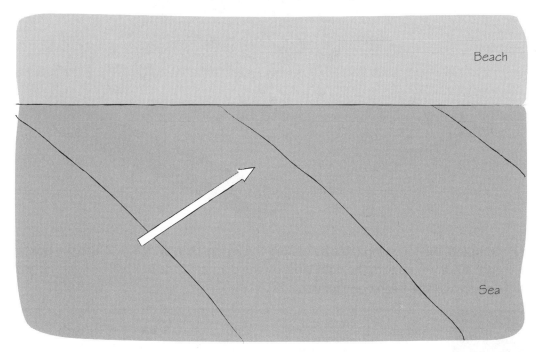

Beach

Sea

*(4 marks)*

> Complete the student's diagram to show the way the swash and backwash move material along the beach.

**(b)** Draw a labelled diagram to illustrate the features of a constructive wave.

*(4 marks)*

> You should also practise drawing and labelling a destructive wave.

# Landforms of coastal erosion

Study this diagram, which shows a range of different landforms of coastal erosion.

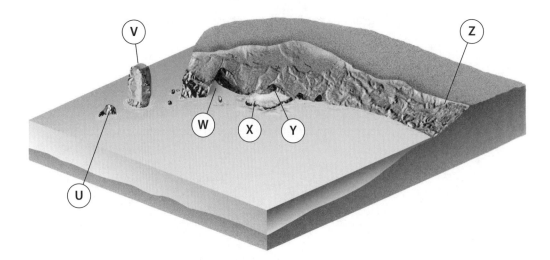

**1 (a)** Complete the following table by writing the letters **U**, **V**, **W**, **X,Y** or **Z** into the boxes to match each landform up with the right label on the diagram. One box will be left empty at the end. One answer has been done for you.

| Landform | Stump | Wave-cut platform | Cliff | Stack | Arch | Wave-cut notch |
|----------|-------|-------------------|-------|-------|------|----------------|
| Letter | U | | | | | |

*(4 marks)*

**(b)** Draw labelled diagram(s) to show how a wave-cut platform is formed.

Guided

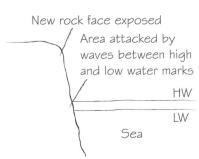

New rock face exposed

Area attacked by waves between high and low water marks

HW

LW

Sea

Complete the student's answer by adding one or more diagrams to show the next stages in the formation of this landform.

*(4 marks)*

# Landforms resulting from deposition

Study this Ordnance Survey map extract showing a coastal landform.

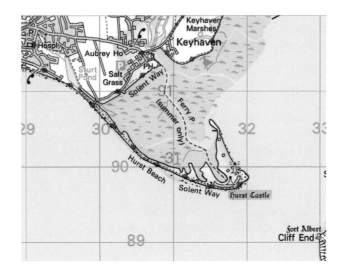

0       1 km

**1 (a)** What type of landform is shown in the map?

..............................................................................................................................................................................

*(1 mark)*

**(b)** Draw labelled diagram(s) to show how this landform is formed.

*(4 marks)*

> Make sure your drawings are clear – it is much more important that the details are clear and accurately labelled than that the drawing is shaded or coloured in. Your labels need to be clear enough to explain the sequence of formation.

# Rising sea levels

tier H

1 Using a **case study** of a coastal area that is threatened by rising sea levels, explain the possible consequences of sea level rise for the people living there.

...............................................................................................................................................

...............................................................................................................................................

...............................................................................................................................................

...............................................................................................................................................

...............................................................................................................................................

...............................................................................................................................................

...............................................................................................................................................

...............................................................................................................................................

...............................................................................................................................................

...............................................................................................................................................

...............................................................................................................................................

...............................................................................................................................................

...............................................................................................................................................

...............................................................................................................................................

...............................................................................................................................................

*(8 marks)*

**Extra space**

...............................................................................................................................................

...............................................................................................................................................

...............................................................................................................................................

...............................................................................................................................................

...............................................................................................................................................

Remember that you will have studied a range of different consequences for your case study which could help you organise your answer: economic, social, environmental and political. Link your points so the answer makes clear arguments and shows the interrelationships between these factors. Using a range of specialist terms for the topic at appropriate points will help towards making an impressive answer.

# Coastal management

Study this map which shows coastal management strategies for the Isle of Wight.

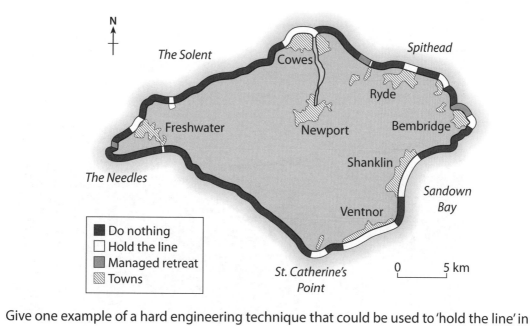

tier **H**

**1** Give one example of a hard engineering technique that could be used to 'hold the line' in the Isle of Wight.

........................................................................................................................................................

*(1 mark)*

tier **F&H**

**2** What are the costs and benefits of a 'do-nothing' approach to coastal management?

........................................................................................................................................................

........................................................................................................................................................

........................................................................................................................................................

........................................................................................................................................................

........................................................................................................................................................

........................................................................................................................................................

........................................................................................................................................................

........................................................................................................................................................

........................................................................................................................................................

........................................................................................................................................................

........................................................................................................................................................

*(6 marks)*

> Remembering both costs and benefits must be covered. Costs can mean more than just financial costs – it can mean a range of disadvantages including environmental, political and social cost.

Case study

# Cliff collapse

Study this photograph, which shows holiday homes on an area of rapid coastal erosion on the Holderness coast.

Guided

1 (a) Describe **two** factors which cause some cliffs, like the ones shown in the photograph, to collapse more quickly along some stretches of coastline than along others.

Factor 1: Coastlines made of resistant rocks are less likely to experience cliff collapse than stretches of coastline made of softer rocks, especially when these are prone to saturation by rainwater ..................................................................................................................

..................................................................................................................................................

..................................................................................................................................................

Factor 2: ..................................................................................................................................

..................................................................................................................................................

..................................................................................................................................................

..................................................................................................................................................

(4 marks)

tier F&H

(b) The photograph shows a holiday home resort threatened by rapid coastal erosion. Describe the impacts that rapid coastal erosion can have on people's lives.

.............................................................................................

.............................................................................................

.............................................................................................

.............................................................................................

.............................................................................................

.............................................................................................

> Costs can mean more than just financial costs – it can mean a range of disadvantages including environmental, political and social costs. You must show how these affect people's lives.

(4 marks)

 **Case study**

# Managing the coast

Study the table below, which shows the costs of constructing and maintaining different strategies of coastal management.

| Strategy | Construction costs (£ per metre) | Maintenance costs (£ per metre per year) |
|---|---|---|
| Rock armour | 1100 | 5 |
| Sea wall | 2000 | 5 |
| Groynes | 2000 | 10 |
| Beach nourishment | 2000 | 2 |

**tier Fn**

**1 (a) (i)** Which is the cheapest strategy to construct?

.................................................................................................................................................

*(1 mark)*

**(ii)** Which is the most expensive strategy to maintain?

.................................................................................................................................................

*(1 mark)*

**tier F&H**

**(b)** Explain how groynes can help to reduce rates of coastal erosion.

.................................................................................................................................................

.................................................................................................................................................

.................................................................................................................................................

.................................................................................................................................................

*(2 marks)*

**tier F&H**

**(c)** Describe one disadvantage of beach nourishment that might mean it is not selected as a method of coastal management in some situations.

.................................................................................................................................................

.................................................................................................................................................

.................................................................................................................................................

.................................................................................................................................................

.................................................................................................................................................

*(2 marks)*

> It is a good idea to revise the advantages **and** disadvantages of a range of coastal management strategies. Cost is often a major concern, but also how often a process has to be repeated to make sure its effectiveness, is also a concern.

# Coastal habitats

**1** Describe the strategies used to conserve a **named** coastal habitat.

**Named example:** .................................................................................................................

.........................................................................................................................................

.........................................................................................................................................

.........................................................................................................................................

.........................................................................................................................................

.........................................................................................................................................

.........................................................................................................................................

.........................................................................................................................................

.........................................................................................................................................

.........................................................................................................................................

.........................................................................................................................................

.........................................................................................................................................

.........................................................................................................................................

.........................................................................................................................................

.........................................................................................................................................

*(8 marks)*

**Extra space**

.........................................................................................................................................

.........................................................................................................................................

.........................................................................................................................................

.........................................................................................................................................

.........................................................................................................................................

> These strategies should be about how the habitat is being protected from the sea (e.g. extended through managed retreat) and how it is being protected from damage by visitors. Make sure your details are specific to your named example.

# Population explosion

Study this table of changes in worldwide birth rates, death rates and infant mortality rates.

| Year | Birth rate per 1000 people | Death rate per 1000 people | Infant mortality rate per 1000 live births |
|------|----------------------------|----------------------------|--------------------------------------------|
| 1800 | 40 | 35 | No data |
| 1850 | 40 | 34 | No data |
| 1900 | 37 | 28 | No data |
| 1950 | 37 | 20 | 126 |
| 2000 | 23 | 9 | 57 |
| 2050 | 14* | 10* | 10* |

*\* Predicted*

**1 (a)** Describe the trends shown in the table.

At the start of the time period shown, death rates are only slightly below birth rates. Natural increase would have been quite low. But as time goes on, the death rate starts to fall while the birth rate .........................................................................

..............................................................................................................................

..............................................................................................................................

..............................................................................................................................

..............................................................................................................................

> Look out for any exceptions to the general trend and include them.

*(4 marks)*

**(b)** Using your own knowledge, explain why the death rate and infant mortality rate shown in the table decreased over time.

..............................................................................................................................

..............................................................................................................................

..............................................................................................................................

..............................................................................................................................

..............................................................................................................................

*(4 marks)*

# Demographic transition

Study this graph representing the changes many countries' populations have gone through over time.

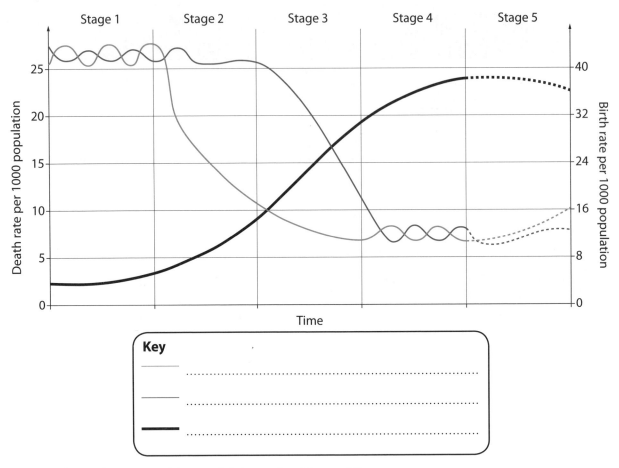

Key

........................................................................

........................................................................

──── ........................................................................

1 (a) What is the name of the model shown by this graph?

.................................................................................................................................

.................................................................................................................................

*(1 mark)*

(b) Select the correct terms from this list to complete the key to the graph. Write your answers into the gaps left for them in the key.

**Birth rate    Life expectancy    Death rate    Natural increase    Total population**

*(3 marks)*

(c) Stage 3 of this model is where many countries in the world are currently located. Which **two** of these countries would you expect to see in this stage? Tick the **two** that you want to select as your answer.

| Countries | Tick to select |
|-----------|----------------|
| Brazil | |
| France | |
| India | |
| UK | |
| USA | |

*(2 marks)*

# Population structure

Study these two population pyramids, one for Afghanistan and one for South Korea.

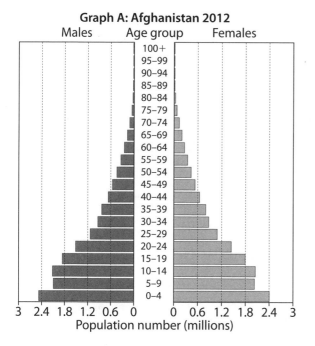

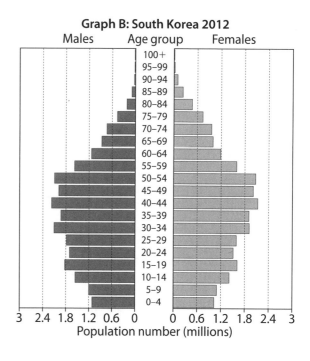

**1 (a)** Graph A shows a country in stage 1 of the Demographic Transition Model. Give **two** features of the shape of the graph that indicate this.

1 ...........................................................................................................

..............................................................................................................

..............................................................................................................

2 ...........................................................................................................

..............................................................................................................

..............................................................................................................

> Remember this question is about the **shape** of the graph. When comparing two graphs check that they are recording things in the same way, for example: the same time period and the same units of measurement.

*(2 marks)*

**(b)** Graph B shows a country in stage 5 of the Demographic Transition Model. Give **two** features of the graph that indicate this.

1 ...........................................................................................................

..............................................................................................................

2 ...........................................................................................................

..............................................................................................................

*(2 marks)*

# Growing pains

Study this diagram of factors that reduce the rate of population growth.

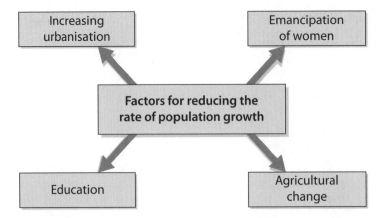

1 Choose **two** of the factors listed or others that you studied. Explain why your chosen factors may reduce the rate of population growth.

Factor 1: The emancipation of women means women having more rights and freedoms in society. When women have more control over their lives they have fewer children.

When a question uses the word 'explain', think of using link words like 'because…' to make sure your answer stays on track.

............................................................................................

............................................................................................

............................................................................................................................................

Factor 2: ..............................................................................................................

............................................................................................................................................

............................................................................................................................................

............................................................................................................................................

*(4 marks)*

# Managing population growth

Study the newspaper article about China's one-child policy.

China's one-child policy was launched in 1980 when leader Deng Xiaoping urged people to have only one child. Deng Xiaoping was worried that if Chinese people had too many babies, the population growth would quickly swamp China's economic development. However, the policy has caused many problems. Families had to pay large fines if they had more than one child. Some families pretended their 'extra' children were nieces or nephews instead of their own children. Because boys are valued more than girls, thousands of female foetuses have been aborted. Over time, this has led to many men not being able to find partners. Because families are restricted to one child, their offspring are very spoiled, leading to the creation of 'Little Emperors'. And, as China's population ages, there is another major problem coming. There will soon be many more old people than young people to care for them.

**tier H**

**Guided**

1   Using the newspaper article and your own knowledge, explain why changes have been made to China's one-child policy since the 1990s.

Although the Chinese government says that the one-child policy has succeeded in preventing 400 million births, the policy has changed since it began in the 1980s because of problems it has caused. In rural areas, people need large families to work the land so the one-child policy has been relaxed in rural areas. Also, parents who are both from a one-child family are allowed to have more than one child because ..........................................................................................

..........................................................................................

..........................................................................................

..........................................................................................

..........................................................................................

..........................................................................................

..........................................................................................

..........................................................................................

..........................................................................................

..........................................................................................

..........................................................................................

..........................................................................................

..........................................................................................

..........................................................................................

> Make sure you read the question carefully. Note that this question says 'since the 1990s'. It is very important that you learn about changes to the policy since the 1990s and apply this to the question, rather than concentrating on an earlier period. Also, make sure you explain the reasons for these changes as well as stating them.

> There are 3 marks available for spellling, punctuation and grammar

*(8 marks)*
*SPaG: 3 marks*

## Case study

# Ageing population

Study this graph which shows the changing population of France, a rich European country. The data are real up to 2012 and then projections for 2022, 2032 and 2042.

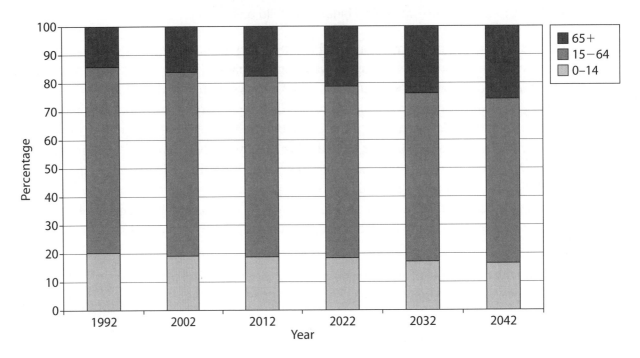

tier F&H

Guided

1   Use the graph to describe what population experts predict will happen to France's population structure in the next 30 years.

The graph shows the proportion of people over 65 years old growing significantly from a figure of around 17 per cent in 2012 to more like 25 per cent in 2042. This continues a trend from 1992. The graph also shows a decline in ................................

........................................................................................................................................

........................................................................................................................................

........................................................................................................................................

........................................................................................................................................

........................................................................................................................................

........................................................................................................................................

........................................................................................................................................

........................................................................................................................................

*(4 marks)*

Make sure you know about different types of graph and understand how to read them.

# Migration: push and pull

Study this photograph, which shows refugees fleeing the conflict in Syria.

 **1 (a)** What is meant by the term 'refugee'?

......................................................................................................................................................................

*(1 mark)*

**Guided** **(b)** Give **two** examples of the sorts of push factor that lead people to become refugees.

Civil wars are often a cause of people to become refugees. Another example is

......................................................................................................................................................................

......................................................................................................................................................................

*(2 marks)*

**(c)** Describe the advantages to the host country of migrants coming from another country to live and work there.

...............................................................................................

...............................................................................................

...............................................................................................

...............................................................................................

...............................................................................................

...............................................................................................

...............................................................................................

......................................................................................................................................................................

......................................................................................................................................................................

......................................................................................................................................................................

> Note that the question is about advantages to the host country, not to the country that the migrants come from. Try to make two points and develop each for a good answer. Also, be careful to take notice of any plurals in a question, e.g. advantages, because it means that more than one has to be covered.

*(6 marks + 3 marks SPaG (F), (6 marks (H))*

# EU movements

Study this newspaper article.

## Immigration to Peterborough

Schools in Peterborough are struggling with the task of teaching so many students who do not speak English as their first language. In Peterborough's primary schools, around 30 per cent of students do not speak English as their first language, 22 per cent in secondary schools – while the national average is 12 per cent. Currently around 2000 new migrants come to Peterborough each year and that can mean up to 900 new students starting in Peterborough's schools at various points in time throughout each school year – the equivalent of three new primary schools!

**1 (a)** Suggest why so many migrants move to places like Peterborough.

> **Guided**

People from poorer EU countries can earn a lot more working (and can send money home) in richer EU countries like the UK than they can at home, and they also .............

.................................................................................................................................

.................................................................................................................................

*(2 marks)*

**(b)** Use the newspaper article and your own knowledge to suggest the impacts of large numbers of migrants moving to a place like Peterborough.

.................................................................................................................................

.................................................................................................................................

.................................................................................................................................

.................................................................................................................................

.................................................................................................................................

.................................................................................................................................

.................................................................................................................................

.................................................................................................................................

.................................................................................................................................

*(4 marks)*

> When you are asked to use stimulus material like this newspaper article in your answer, do not just repeat what it says. You need to adapt the information it gives to build your answer to the question. Think of using links like 'which means that…'. A good answer for this question would consider both positive and negative impacts.

# Urbanisation goes global

Study this table, which shows the percentage of the total population that is urban in selected countries between 1985 and 2010.

**Table 1** Urban population as a percentage of total population

|  | 1985 | 1995 | 2010 |
|---|---|---|---|
| Brazil | 71% | 78% | 87% |
| Botswana | 27% | 49% | 61% |
| China | 23% | 31% | 45% |
| Malaysia | 46% | 56% | 72% |
| UK | 89% | 89% | 90% |

*Source: World Bank data.*

1　**(a)** Use the data provided in **Table 1** to complete the following table. Write the names of the countries into the last column of the table. The first one is completed for you.

| Description of urbanisation | Name of country |
|---|---|
| This country shows both a rapid rate of urbanisation and over 80% urban population | Brazil |
| This country's urbanisation rate has slowed to almost nothing | |
| This country saw more rapid urbanisation from 1985 to 1995 than from 1995 to 2010 | |
| This Asian country still has a majority rural population | |

*(3 marks)*

▷ **Guided** ▷　**(b)** Name **two** factors that help explain rapid urbanisation in poorer parts of the world.

High rates of rural–urban migration and ..................................................................

...................................................................................................................

...................................................................................................................

*(2 marks)*

> The first factor has been added already, now complete the answer by writing in a second factor.

# Inner city issues

Study this photograph of part of a city in the UK.

Photos like this contain a great deal of information so study them carefully for the clues you need to answer the question.

**1 (a)** Name the part of the city shown in the photogaph.

Guided    The inner city. ................................................................................................................................

*(1 mark)*

**(b)** Describe **one** piece of evidence from the photograph for your choice of answer to **1(a)**.

...........................................................................................................................................................

...........................................................................................................................................................

...........................................................................................................................................................

...........................................................................................................................................................

*(2 marks)*

**(c)** Explain why areas such as the one shown in the photograph may need to be revitalised.

...........................................................................................................................................................

...........................................................................................................................................................

...........................................................................................................................................................

...........................................................................................................................................................

...........................................................................................................................................................

...........................................................................................................................................................

...........................................................................................................................................................

...........................................................................................................................................................

...........................................................................................................................................................

*(4 marks)*

# Housing issues and solutions

Study this photograph, which shows the Athletes' Village accommodation blocks built for the 2012 London Olympics.

**1 (a)** Which of the following options would help tackle the housing problems that face local people from this inner city area? Tick the box to show your choice.

| Option | Tick to select |
| --- | --- |
| Make the Athletes' Village into office blocks. | |
| Make the Athletes' Village into luxury flats for wealthy city workers. | |
| Make the Athletes' Village into affordable homes for local people. | |

*(1 mark)*

**(b)** Explain why housing often causes problems for people living in the inner city.

The inner city has mostly old housing, often split up into small flats, and tower blocks built in the 1960s which were badly designed and not very well built, so ........................

...............................................................................................................................

...............................................................................................................................

...............................................................................................................................

...............................................................................................................................

...............................................................................................................................

...............................................................................................................................

...............................................................................................................................

...............................................................................................................................

...............................................................................................................................

...............................................................................................................................

*(6 marks + 3 marks SPaG (F), (6 marks (H))*

# Inner city challenges

Study this photograph, which shows the site of the London 2012 Olympics at the start of development. This was a brownfield site in part of London's inner city.

> There are extra marks available for spelling, punctuation and grammar. Make sure you check each of those and that your answer is well organised.

**1** Explain the advantages and disadvantages of using a brownfield site for the Olympics.

Guided

Using a previously run-down area of London for the Olympics meant the whole area was regenerated, not just for the Olympic Games themselves but for a long time after. Also, .............................................................................................................................

.........................................................................................................................................................

.........................................................................................................................................................

.........................................................................................................................................................

.........................................................................................................................................................

.........................................................................................................................................................

.........................................................................................................................................................

.........................................................................................................................................................

.........................................................................................................................................................

.........................................................................................................................................................

.........................................................................................................................................................

.........................................................................................................................................................

*(8 marks)*
*SPaG: 3 marks*

**Extra space**

.........................................................................................................................................................

.........................................................................................................................................................

.........................................................................................................................................................

# Squatter settlements

Read the following extract, which describes the health and social problems found in many squatter settlements.

> Life in squatter settlements often faces significant health risks. There is rarely good sanitation (sewage systems). Drinking water is not often safe. Food storage facilities are usually poor. There is little protection from mosquitoes, fleas and ticks. This means that people living in squatter settlements are exposed to a wide range of diseases. Because most people do not have access to electricity, people cook with wood and there is a lot of air pollution inside the houses. And because people are so crowded together there is a lot of stress, violence and social problems such as drugs. There is a particular risk for young babies and infant mortality rates in squatter settlements are often high.

**1 (a)** Using the extract and your own knowledge, describe the social problems faced by people living in squatter settlements.

> The extract describes health and social problems. The health problems are linked to there often not being any clean water to drink and often no clean way of removing human waste, which is usually dumped into nearby water sources. The social problems

........................................................................................................................

........................................................................................................................

........................................................................................................................

........................................................................................................................

........................................................................................................................

........................................................................................................................

........................................................................................................................

*(4 marks)*

**(b)** Using your own knowledge, explain **two** reasons why people prefer to live in squatter settlements than go home to the countryside.

1 ......................................................................................................................

........................................................................................................................

........................................................................................................................

........................................................................................................................

........................................................................................................................

2 ......................................................................................................................

........................................................................................................................

........................................................................................................................

........................................................................................................................

*(4 marks)*

# Squatter settlement redevelopment

**1** The following is a list of some strategies that have been used to tackle the problems of squatter settlements.

> • Using bulldozers to demolish squatter settlements and rehouse residents in new flats.
>
> • Providing building materials, technical help and loans so residents can improve the settlement for themselves.
>
> • Site and service – local authorities provide a site and basic services for residents to build their own homes.

Choose **two** of the solutions listed or others that you have studied.

Explain how well your chosen solutions might deal with the problems of squatter settlements.

Solution 1 ..............................................................................................................................................

........................................................................................................................................................

........................................................................................................................................................

........................................................................................................................................................

........................................................................................................................................................

........................................................................................................................................................

........................................................................................................................................................

Solution 2 ..............................................................................................................................................

........................................................................................................................................................

........................................................................................................................................................

........................................................................................................................................................

........................................................................................................................................................

........................................................................................................................................................

*(6 marks + 3 marks SPaG (F), (6 marks (H))*

**Extra space**

.......................................................................................................

.......................................................................................................

.......................................................................................................

.......................................................................................................

.......................................................................................................

> Because the question asks 'how well', you should look at the advantages and disadvantages of the solutions. The question does not ask for a case study but use one if it helps you decide the advantages and disadvantages.

# Rapid urbanisation and the environment

**1 (a)** Complete the sentences below to describe some of the ways one city is attempting to manage growing air and water pollution. Choose the correct words from the following list.

**green belt    declining    air    blocking    bypass    repairing    ~~growing~~    water**

Cairo is a city in Egypt. It has a rapidly ␣growing␣ population. There are major problems with air

pollution from traffic congestion and ............................................. pollution from sewage,

industrial pollution and waste being dumped in rivers and streams. Traffic congestion is being

addressed by the new Cairo metro and the building of a new ............................................. round

Cairo. Sewage on the streets is being dealt with by the Greater Cairo Waste Water Project, which is

............................................. and upgrading sewers across Cairo.

*(4 marks)*

> The first gap has been filled. Now work out the correct terms to use to complete the rest of the paragraph.

**(b)** Give **two** effects of water pollution in urban areas.

..................................................................................................................................................

..................................................................................................................................................

..................................................................................................................................................

*(2 marks)*

**(c)** Give **two** effects of air pollution in urban areas.

..................................................................................................................................................

..................................................................................................................................................

..................................................................................................................................................

..................................................................................................................................................

*(2 marks)*

> For 'give' questions like these, you do not need to write full statements. You can just write down your response, e.g. 'Causes health problems for people with respiratory illnesses'.

# Sustainable cities

Study this photograph, which shows a landfill waste disposal site in the UK.

**1 (a)** 'Up to 90 per cent of household waste is recyclable'. What is meant by the term 'recyclable'?

.................................................................................................................................................

.................................................................................................................................................

*(1 mark)*

**Guided**

**(b)** Explain why landfill is not a sustainable way to deal with urban waste.

The UK has currently almost run out of sites that are suitable for landfill, so ...............

.................................................................................................................................................

.................................................................................................................................................

.................................................................................................................................................

.................................................................................................................................................

.................................................................................................................................................

.................................................................................................................................................

*(2 marks)*

# The rural-urban fringe

Study this aerial photograph of the Bluewater shopping centre in Kent.

**1 (a)** The following statements are about pressures on the rural–urban fringe.

Complete the table by ticking the correct box to show whether each statement is **True** or **False**. The first answer is completed for you.

| | Statement | True | False |
|---|---|---|---|
| **A** | Bluewater shopping centre is built in an urban area. | | ✔ |
| **B** | One of the attractions of this site for Bluewater was the access to good transport links. | | |
| **C** | One of the attractions of the site was to encourage people to shop in the town centre. | | |
| **D** | As a result of Bluewater being built here, this rural area is more built up and crowded than before. | | |

*(3 marks)*

**(b)** Out-of-town shopping centres put pressure on the rural–urban fringe. Give **one** other example of a pressure on the rural–urban fringe.

.................................................................................................................................................

*(1 mark)*

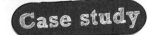 Case study

# Rural depopulation and decline

Study this information based on 2012 figures.

- Herefordshire is a predominantly rural area.
- The average wage in England is £21 560 per year.
- The average wage in urban areas of England is £23 560 per year.
- The average wage in Herefordshire is £16 398.

 tier F&H

**1 (a)** Explain how this information could help account for rural depopulation in Herefordshire.

..................................................................................................................................................

..................................................................................................................................................

*(2 marks)*

tier Fn

**(b)** The following is a list of other factors which may increase depopulation in remote rural areas.

- Lack of affordable housing to buy or rent.
- Lack of broadband access or mobile phone coverage.
- Very little help for parents with young children.

Choose **two** of the factors listed.

Explain why your chosen factors may increase rural depopulation.

Factor 1 ........................................................................................................................................

..................................................................................................................................................

..................................................................................................................................................

..................................................................................................................................................

..................................................................................................................................................

Factor 2 ........................................................................................................................................

..................................................................................................................................................

..................................................................................................................................................

..................................................................................................................................................

..................................................................................................................................................

*(4 marks)*

> You should make one developed point per factor.

# Supporting rural areas

Study this press report about a government initiative in Cumbria, a rural area in northern England.

> The Department for Environment, Food and Rural Affairs has announced a major investment to support rural life in Cumbria. The new Rural Growth Network will:
>
> - Create hubs for rural businesses in 11 sites across Cumbria. These will provide business skills and support services for small businesses.
>
> - Provide a rural web portal dedicated to linking up rural businesses and providing support.
>
> - Give a special boost to industries connected to food and drink, digital and creative industries, agribusiness and forestry and adventure and sports activities.
>
> - Special funding for enterprises run by women.
>
> The government expects this funding to create 480 new business start-ups and 900 jobs.

1  **(a)** Which of the following is this government initiative designed to do?

Is it:

- to conserve farm resources

- to protect the rural environment

- to support the rural economy?

...................................................................................................................................................

*(1 mark)*

**(b)** Explain why the government is keen to encourage other industries apart from farming in this rural area.

Farming is very important in rural areas but modern farming uses lots of machinery and does not employ many people. Also the jobs it creates are not always very well paid and do not suit everyone. So ...............................................................................

...................................................................................................................................................

...................................................................................................................................................

...................................................................................................................................................

...................................................................................................................................................

...................................................................................................................................................

...................................................................................................................................................

...................................................................................................................................................

...................................................................................................................................................

*(4 marks)*

> One developed point has already been made to start this answer off; you should add another one explaining why other industries can help widen the range of jobs available, tapping into better paid lines of work.

 **Case study** Commercial farming 1

**Guided**

**1** **(a)** Describe the features that make an agribusiness different from a small family-run farm.

Agri-businesses are run on a very large scale. Family-run farms are not as large – in fact agribusiness farm holdings are generally made up of several old family farms joined together ............................................................................................................

..............................................................................................................................

..............................................................................................................................

..............................................................................................................................

..............................................................................................................................

..............................................................................................................................

..............................................................................................................................

*(4 marks)*

**(b)** Explain how farming in a **named area** of the UK is affected by the demands of supermarkets and food processing companies.

..............................................................................................................................

..............................................................................................................................

..............................................................................................................................

..............................................................................................................................

..............................................................................................................................

..............................................................................................................................

..............................................................................................................................

..............................................................................................................................

..............................................................................................................................

..............................................................................................................................

..............................................................................................................................

..............................................................................................................................

..............................................................................................................................

*(8 marks)*
*SPaG: 3 marks*

> Remember to give the name of your case study area and make sure you deal with the effects of **both** supermarkets and of food processing companies and the standards and prices that they demand from their suppliers. There are also 3 marks available for spelling, punctuation and grammar so make sure you check each of these and that your answer is well organised.

**Extra space**

..............................................................................................................................

..............................................................................................................................

..............................................................................................................................

**CHANGING RURAL ENVIRONMENTS**

# Commercial farming 2

Study the chart, which shows reasons why people in the UK choose to buy organic food.

**The top reasons for buying organic**

Fewer chemicals
62%

Natural and unprocessed
57%

Healthier for me and my family
52%

Better for nature/the environment
47%

Organic food tastes better
44%

Safer to eat
41%

Organic farming has high animal welfare
34%

More ethical
33%

No GM ingredients
29%

There are good offers
13%

**1 (a)** What is meant by the term 'GM ingredients'?

..............................................................................................................................

..............................................................................................................................

..............................................................................................................................

*(2 marks)*

**(b)** Using the chart, explain the attractions of organic food to some UK customers.

> There are two main reasons in the chart about why people like organic food. The first is that organic food seems safer and healthier to eat because it does not contain chemicals or GM ingredients. The second reason is ....................................................

..............................................................................................................................

..............................................................................................................................

..............................................................................................................................

..............................................................................................................................

..............................................................................................................................

..............................................................................................................................

*(4 marks)*

> Try to develop the information you take from the graph.

# Changing rural areas: tropical 1

Study this photograph, which shows Guarani people working on a sugar cane plantation in Brazil. The Guarani are the indigenous inhabitants of huge areas of South America, including Brazil. Sugarcane is an agri-business crop, grown to convert into bio-fuel.

 tier F&H

**1** **(a)** Is sugar cane grown to feed local families or as a cash crop? Tick the answer you think is the correct one:

☐  to feed local families

☐  as a cash crop.  *(1 mark)*

▷ **Guided**

**(b)** Describe the main impacts that cash crop farming has on traditional farming in tropical rural areas.

Because cash crop farming takes up a lot of land, there is less farming land for traditional farmers – traditional farming uses up a lot of land too because small plots in the forest are only cultivated for a couple of years before being left to recover. The second impact is ...............................................

> Try to make two developed points for this sort of question.

...........................................................................................................

...........................................................................................................

...........................................................................................................

...........................................................................................................

...........................................................................................................

...........................................................................................................

...........................................................................................................

*(6 marks + 3 marks SPaG (F), 6 marks (H))*

# Changing rural areas: tropical 2

Study the following table that shows the impact of building appropriate technology water projects in a village in Gujarat, India.

| Thunthi Kankasiya village | Before 1991 | By 2000 |
|---|---|---|
| All-year drinking water wells | 0 | 23 |
| River dams | 0 | 1 |
| Months of water availability | 4 | 12 |
| Land under cultivation (hectares) | 85 | 135 |
| Number of crops grown through the year | 0–1 | 2–3 |
| Agricultural production (quintals per hectare) | 900 | 4000 |
| Migration rate from the village (15–40 year olds) | 78% | 5% |
| Income per household (rupees per year) | 8600 | 35 600 |

A hectare is a unit of land measurement. It is a square of land with sides each 100 metres long: 10 000 m². A quintal is a measurement of mass. In India one quintal is 100 kilograms.

**1 (a)** Complete the following paragraph about the data in the table above, by choosing the correct words from the following list.

> **twelve    63    well    73    increased    dam    six    decreased**

Building a river ................... in Thunti Kankasiya village has had a big impact. Instead of water only being available for four months, water is available for ............... months. The amount of land farmed, crop production and farmers' incomes have all .................... Migration from the village has decreased by .............. per cent.

*(4 marks)*

**Guided**

**(b)** Describe **two** impacts of high rural–urban migration on villages like Thunthi Kankasiya.

One impact is that it is generally younger men who leave the village and that makes it harder for the women, children and old men who remain to farm the land and increase food production. The second impact is ................................................................................................

.............................................................................................................................

.............................................................................................................................

.............................................................................................................................

.............................................................................................................................

.............................................................................................................................

*(4 marks)*

# Measuring development 1

Study this scatter graph comparing infant mortality with GDP per head (a similar measure to GNI per head).

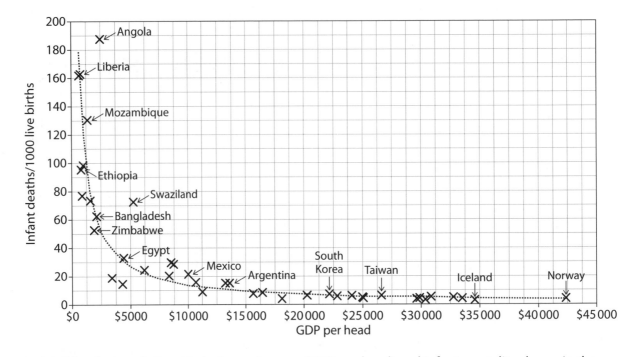

**1 (a)** What is the relationship between income (GDP per head) and infant mortality shown in the scatter graph?

......................................................................................................................................................

......................................................................................................................................................

*(1 mark)*

**(b)** Give **two** examples of the limitations of using a single development measure to indicate how developed a country is.

Example 1:...............................................................................................................................

......................................................................................................................................................

Example 2:...............................................................................................................................

......................................................................................................................................................

*(2 marks)*

> By 'relationship', this question is asking about any link between the two factors.

# Measuring development 2

Study this map classifying countries of the world according to their Human Development Index score.

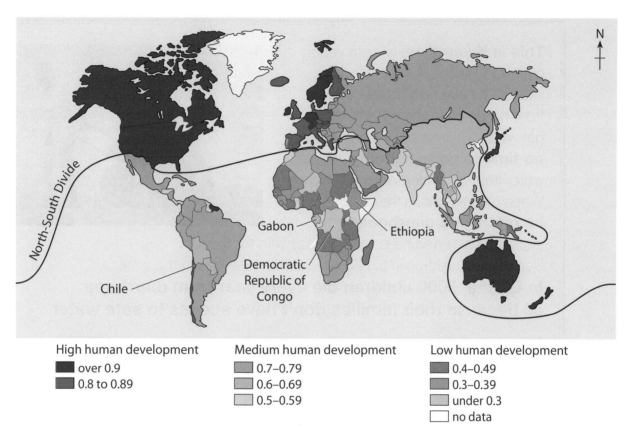

High human development
- ■ over 0.9
- ■ 0.8 to 0.89

Medium human development
- ■ 0.7–0.79
- ■ 0.6–0.69
- ■ 0.5–0.59

Low human development
- ■ 0.4–0.49
- ■ 0.3–0.39
- ■ under 0.3
- ☐ no data

**1 (a)** Use the map and key to complete the following table. Some of the table has been completed for you.

| Country | HDI classification | North or South? |
|---|---|---|
| Democratic Republic of Congo | Low human development | South |
| Ethiopia | | South |
| Gabon | | South |
| Chile | | South |

*(3 marks)*

**(b)** Explain why the classification in the map is superior to classifying the world into rich North and poor South.

.................................................................................................................................................

.................................................................................................................................................

.................................................................................................................................................

.................................................................................................................................................

*(2 marks)*

Had a go ☐    Nearly there ☐    Nailed it! ☐

# What causes inequalities?

Study this charity advert about the importance of access to safe water.

**This is Kwame.** He is eight years old and lives in Ghana. In the dry season, it takes two hours to walk from his home to the waterhole to get water. That means **Kwame has no time to go to school**. And the water from the waterhole is not clean. When Kwame drinks the water, **he often gets diarrhoea** and stomach pains and he is too ill to play with his friends. Sometimes, children in his village die from diseases in the water.
**In Ghana, 5000 children die every year from diarrhoea: all because their families don't have access to safe water.**

**1** Use the advert and your own knowledge to explain how access to water can affect people's standards of living.

> The question mentions access to water so there are two aspects to deal with here: access to enough water and access to water that is safe to drink. Quantity and quality.

....................................................................................

....................................................................................

....................................................................................

....................................................................................

....................................................................................

....................................................................................

....................................................................................

....................................................................................

....................................................................................

....................................................................................

....................................................................................

....................................................................................

*(6 marks + 3 marks SPaG (F), 6 marks (H))*

**Extra space**

> There are 3 marks available for spelling, punctuation and grammar. Make sure you check each of these and that your answer is well organised.

....................................................................................

....................................................................................

....................................................................................

Case study

# The impact of
# a natural hazard

Study this photograph taken in Haiti after the 2010 earthquake.

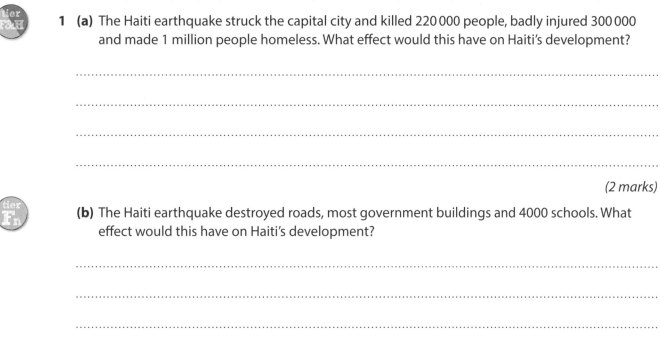

tier
F&H

**1 (a)** The Haiti earthquake struck the capital city and killed 220 000 people, badly injured 300 000 and made 1 million people homeless. What effect would this have on Haiti's development?

..................................................................................................................................................................

..................................................................................................................................................................

..................................................................................................................................................................

..................................................................................................................................................................

*(2 marks)*

tier
Fn

**(b)** The Haiti earthquake destroyed roads, most government buildings and 4000 schools. What effect would this have on Haiti's development?

..................................................................................................................................................................

..................................................................................................................................................................

..................................................................................................................................................................

..................................................................................................................................................................

*(2 marks)*

# Is trade fair?

Study this graph of Fair Trade banana sales in the UK.

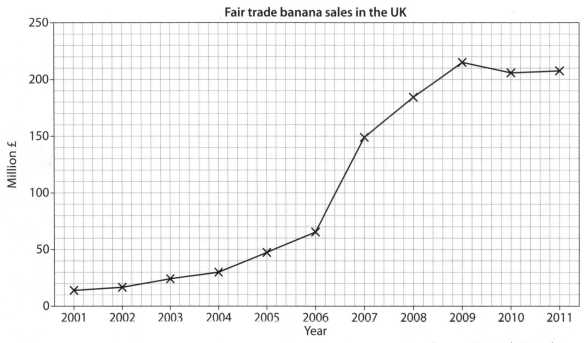

Source: Fairtrade Foundation

1  (a)  By how much have sales of Fair Trade bananas in the UK increased by between 2001 and 2011?

.................................................................................................

.................................................................................................

*(1 mark)*

> You need to be as accurate as possible in your answers to graph questions. Use a ruler or a piece of paper to read off the exact values.

(b)  Explain how Fair Trade schemes try to reduce trade problems faced by poorer countries.

.................................................................................................

.................................................................................................

.................................................................................................

.................................................................................................

.................................................................................................

.................................................................................................

.................................................................................................

.................................................................................................

.................................................................................................

.................................................................................................

.................................................................................................

> Your answer needs to identify trade problems faced by poorer countries, and then also how Fair Trade schemes try to reduce these.

*(4 marks)*

# Case study **Aid and development**

Study this map of the global distribution of international aid.

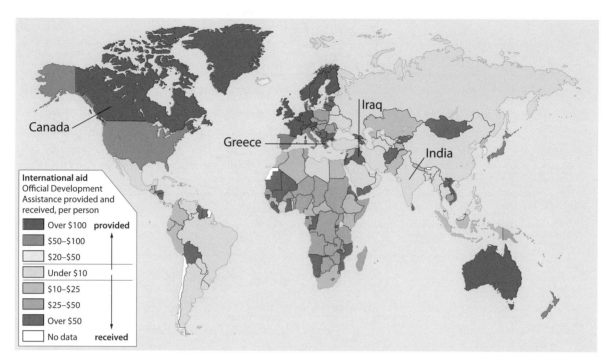

**1 (a)** Complete the table below by ticking the correct box to show whether each country is a provider of aid, a receiver of aid, or both a provider and a receiver.

| Country | Provider of aid | Receiver of aid | Both a provider and a receiver |
|---------|-----------------|-----------------|--------------------------------|
| Canada  |                 |                 |                                |
| Greece  |                 |                 |                                |
| India   |                 |                 |                                |
| Iraq    |                 |                 |                                |

*(4 marks)*

**(b)** Describe the distribution of the countries that donate the most and the countries that receive the most as shown in the map.

The countries that donate the most are all high-income, industrialised countries in western Europe, North America and Australia. ...............................................................

...............................................................

> Two points with development of each would provide a good answer to this question.

..............................................................................................................

..............................................................................................................

..............................................................................................................

..............................................................................................................

..............................................................................................................

*(4 marks)*

Had a go ☐    Nearly there ☐    Nailed it! ☐

**Case study** Development and the EU

**tier H**

1   Describe **one** or **more** policies used by the European Union to attempt to reduce the different levels of development between countries in the EU.

..................................................................................................................................

..................................................................................................................................

..................................................................................................................................

..................................................................................................................................

..................................................................................................................................

..................................................................................................................................

..................................................................................................................................

..................................................................................................................................

..................................................................................................................................

..................................................................................................................................

..................................................................................................................................

..................................................................................................................................

..................................................................................................................................

..................................................................................................................................

*(8 marks)*
*SPaG: 3 marks*

**Extra space**

..........................................................................................................

..........................................................................................................

..........................................................................................................

..........................................................................................................

..........................................................................................................

..........................................................................................................

..........................................................................................................

..........................................................................................................

..........................................................................................................

Try to give specific details, link statements together and identify interrelationships between factors to show how each of the policies aim to reduce the development gap.

Remember that there are 3 marks available for spelling, punctuation and grammar so make sure you check each of these and that your answer is clear and well organised.

# Going global

Study this extract from an article about call centres in the Philippines.

> In 2011, the Philippines became the country with the highest number of call centre employees – 600 000. For millions of customers in Western countries like the UK and USA, that meant calls to sort out a software glitch, make an insurance claim or book a ticket to a show were handled by a Filipino.
>
> Wages are low in the Philippines and many in the country still live in poverty. A job in a call centre is a good job to get. Most Filipinos speak English in an accent that is easy for US and UK customers to understand. And call centre companies in the Philippines have invested a lot of money in improving their services. They want customers to get a better service when they contact a call centre in the Philippines. And they want to be able to respond to customers who prefer to email, message or get help via social media.

**1** **(a)** Use the extract about call centres in the Philippines to help you complete this paragraph. Choose the correct words from the list to fill in the gaps.

**cheaper    loudly    interdependence    English    exported    invested**

Companies with customers in countries like the UK and US use call centres in the Philippines

because they are much ................................ than call centres in the UK or US. Most Filipinos also

speak ................................ with an accent that people in the US and UK find easy to understand.

Customers want to access help by email, messaging and via social networks as well as phone calls,

so it is important that Filipino companies have ................................ a lot in new technology and

in improving services.

*(3 marks)*

**(b)** Give **one** example of a technological development that has made the development of call centres abroad possible.

.................................................................................................................................................

*(1 mark)*

> For a question like this, you can just write down the name of your example, e.g. 'internet telephony', you **don't** need to waste time saying 'One example of a technological development that has made the development of call centres abroad possible is Internet telephony'.

# TNCs

**1** Use a case study to describe the advantages and disadvantages to a poorer country of a TNC locating branches or factories there.

.................................................................................................

.................................................................................................

.................................................................................................

.................................................................................................

.................................................................................................

.................................................................................................

.................................................................................................

.................................................................................................

.................................................................................................

.................................................................................................................

.................................................................................................................

.................................................................................................................

.................................................................................................................

.................................................................................................................

> Make sure you read the question carefully. Here, you are told to use a case study in your answer, so make sure you include details from the one you've studied. Make sure you cover advantages **and** disadvantages for the country in question and try to link your points together.

*(8 marks)*
*SPaG: 3 marks*

**Extra space**

.................................................................................................................

.................................................................................................................

.................................................................................................................

.................................................................................................................

.................................................................................................................

.................................................................................................................

.................................................................................................................

> Remember there are also 3 marks available for spelling, punctuation and grammar. Make sure you check each of these and that your answer is clear and well organised.

# Manufacturing changes

1  The following is a list of some factors which may affect where manufacturing is located in the world in the future.

> • Industrialisation in NICs
>
> • De-industrialisation in HICs
>
> • Technological advances in manufacturing

Choose **two** of the factors listed.

Explain why your chosen factors may affect future manufacturing locations around the world.

Factor 1: ............................................................................................................................................

....................................................................................................................................................................

....................................................................................................................................................................

....................................................................................................................................................................

....................................................................................................................................................................

Factor 2: ............................................................................................................................................

....................................................................................................................................................................

....................................................................................................................................................................

....................................................................................................................................................................

....................................................................................................................................................................

*(4 marks)*

**Extra space**

....................................................................................................................................................................

....................................................................................................................................................................

....................................................................................................................................................................

....................................................................................................................................................................

....................................................................................................................................................................

....................................................................................................................................................................

> Even though the question lists three factors, you only need to choose two. Make sure you read instructions like these very carefully.

# China

1  Explain why China's industrialisation has been so rapid.

..................................................................................................................................

..................................................................................................................................

..................................................................................................................................

..................................................................................................................................

..................................................................................................................................

..................................................................................................................................

..................................................................................................................................

..................................................................................................................................

..................................................................................................................................

..................................................................................................................................

..................................................................................................................................

*(6 marks + 3 marks SPaG (F), 6 marks (H))*

**Extra space**

..................................................................................................................................

..................................................................................................................................

..................................................................................................................................

..................................................................................................................................

..................................................................................................................................

..................................................................................................................................

..................................................................................................................................

..................................................................................................................................

> Try to provide details and relevant examples from your case study to back up your answer.

> Remember, there are also marks available for SPaG so check your spelling, punctuation and grammar is correct and you have used relevant geographical terminology where you can.

# More energy!

Study this map showing the percentage by which total energy consumption has increased in different parts of the world between 2010 and 2011.

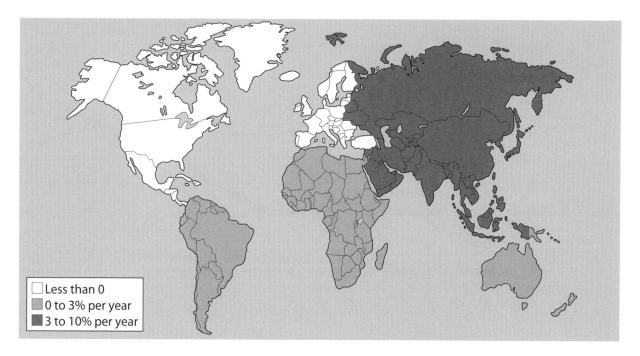

Less than 0
0 to 3% per year
3 to 10% per year

1  Complete the table by ticking the correct box to show whether each of these statements about the map is **True** or **False**.

| Statement | True | False |
|---|---|---|
| Asia's energy consumption has increased by more than 10 per cent between 2010 and 2011. | | |
| There is no increase shown for sub-Saharan Africa. | | |
| Generally, the highly industrialised countries have the lowest percentage increases. | | |
| The area of the biggest increase includes China and India. | | |

*(4 marks)*

2  Which continent showed the greatest increase in the consumption of energy between 2010 and 2011?

...................................................................................................................................

*(1 mark)*

Case study **Sustainable energy use**

Study this graph showing how carbon emissions from consumption of energy have changed for four countries between 1992 and 2010.

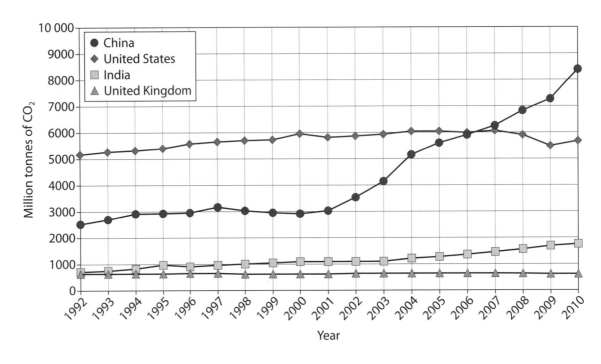

1 **(a)** Which country was emitting the most carbon from energy consumption in 2001?

...................................................................................................................................................................

*(1 mark)*

**(b)** Describe the changes over time shown in the graph.

...................................................................................................................................................................

...................................................................................................................................................................

...................................................................................................................................................................

...................................................................................................................................................................

...................................................................................................................................................................

...................................................................................................................................................................

...................................................................................................................................................................

...................................................................................................................................................................

...................................................................................................................................................................

...................................................................................................................................................................

*(4 marks)*

**(c)** In which year did China become the country which produced the most carbon emissions?

.......................................................................................................

*(1 mark)*

Use a ruler to get accurate readings from graphs. Use accurate figures in your description, not just vague statements like 'it went up a lot'. Try to also refer to all the countries.

# Food: we all want more

Study this diagram explaining the repercussions of increased demand for food in some areas of both poorer and richer countries.

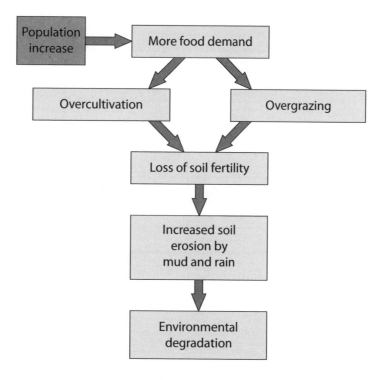

**1 (a)** Why does population increase lead to increasing demand for food?

..................................................................................................................................................

*(1 mark)*

**(b)** Describe **one** other effect of the increasing demand for food.

..................................................................................................................................................

..................................................................................................................................................

*(2 marks)*

**(c)** Using the diagram and your own knowledge, explain the implications of farmers using marginal land to help meet an increasing demand for food.

..................................................................................................................................................

..................................................................................................................................................

..................................................................................................................................................

.................................................................................................

.................................................................................................

.................................................................................................

.................................................................................................

.................................................................................................

*(4 marks)*

> Try not to only repeat what is in the diagram. When a question asks you to use the stimulus resource **and** your own knowledge, the 'your own knowledge' bit could mean adding extra points, using your own knowledge to interpret the stimulus resource, or adding examples to support what the resource shows.

# The tourism explosion

Study this table, which shows the numbers of international tourists according to the world region that they travelled from for the years 1990, 2000 and 2010. The final column shows what percentage share of all tourist arrivals each world region had in the year 2010.

| | International tourist arrivals | | | Share (%) |
|---|---|---|---|---|
| | 1990 | 2000 | 2010 | 2010 |
| **World** | **435 million** | **675 million** | **940 million** | **100%** |
| From: | | | | |
| Europe | 262 million | 385 million | 475 million | 51% |
| Asia and the Pacific | 56 million | 110 million | 204 million | 22% |
| Americas | 93 million | 128 million | 151 million | 16% |
| Middle East | 10 million | 24 million | 60 million | 6% |
| Africa | 15 million | 27 million | 50 million | 5% |

*Source: World Tourism Organization (UNWTO)*

**tier Fn**

1  (a) How many tourists from Europe travelled to another country in 2010?

.................................................................................................................................................................

*(1 mark)*

(b) How many more tourists from the Asia and the Pacific region travelled to another country in 2010 compared with 1990?

.................................................................................................................................................................

*(1 mark)*

**tier Fn**

**Guided**

2  Three main factors for the growth in global tourism are stated below. Briefly explain why each one is important.

(a) Greater wealth: As people in a country earn more money and have more disposable income, more of them can afford to have a foreign holiday. ..........................................

.................................................................................................................................................................

(b) More leisure time:

.................................................................................................................................................................

(c) Cheaper travel:

.................................................................................................................................................................

*(3 marks)*

> A question like this just needs a brief explanation for each factor.

# Tourism in the UK

Study the following table, which shows how many holiday trips were taken by people in selected EU countries in 2011 and whether they went on holiday in their own country or went to another country.

| Country | Total number of holiday trips | % of trips that were in home country | % of trips that were to another country |
|---|---|---|---|
| Germany | 215 million | 66 | 34 |
| Spain | 122 million | 92 | 8 |
| The Netherlands | 30 million | 48 | 52 |
| UK | 117 million | 62 | 38 |

**1 (a)** According to the table, which country of the four had the most holiday trips in 2011?

..........................................................................................................................................................

*(1 mark)*

**(b)** Which country's residents were the most likely to have holiday trips in their own country?

..........................................................................................................................................................

*(1 mark)*

**2** Describe **two** factors that might make people decide to take their holiday in their own country rather than abroad.

Factor 1: If there was a really big event in their own country, like the Olympics and Paralympics, people might decide to visit some of the events. ..........................................................

Factor 2: ...............................................................................................................................................

..........................................................................................................................................................

..........................................................................................................................................................

..........................................................................................................................................................

..........................................................................................................................................................

*(4 marks)*

> The first of these two factors has been completed for you – now do the second factor yourself.

**Case study**

# UK tourism: coastal resort

Study this diagram, which shows different stages in a model often used in geography.

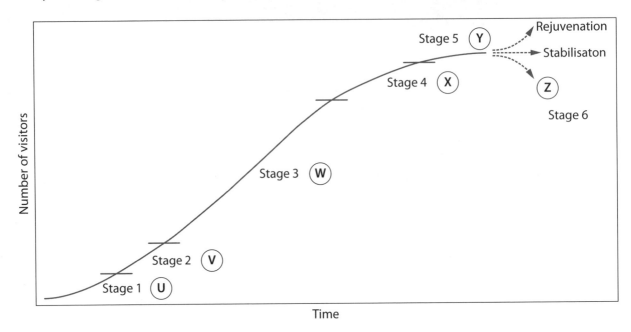

**1  (a)** What is the name of the model shown in the diagram? **Tick the correct box**.

Demographic Transition Model   ☐

Tourist area life cycle model      ☐                                          *(1 mark)*

**(b)** The letters **U** to **Z** refer to stages in the model shown in the diagram.

Complete the table below by writing **one** of the letters **U** to **Z** against each correct stage in the model. The stages have been jumbled up. Two have been completed for you.

| Stage | Letter |
|---|---|
| **Stagnation** – visitor numbers peak, the numbers of tourists start to make the area less attractive to new visitors. | Y |
| **Decline** – the area becomes less attractive and visitor numbers drop. | Z |
| **Exploration** – a few people start to visit the area, before it has many tourism facilities. | |
| **Consolidation** – visitor numbers are getting near their peak. | |
| **Development** – the number of visitors booms as tourism becomes really big business. | |
| **Involvement** – visitor numbers begin to grow as local people start to provide more tourist facilities. | |

*(3 marks)*

# UK tourism: National Park

**1** Choose **either** a National Park **or** a coastal resort in the UK.

**Name of National Park or coastal resort**: ...................................................................................

Explain why the area you have chosen attracts large numbers of tourists each year.

............................................................................................

............................................................................................

............................................................................................

............................................................................................

............................................................................................

............................................................................................

............................................................................................

............................................................................................

| Linking your points together and using some relevant details and examples is very important. Try to use geographical terms and remember that there are 3 marks available for spelling, punctuation and grammar, so make sure you check each of these and that your answer is clear. |
| --- |

..................................................................................................................................

..................................................................................................................................

..................................................................................................................................

*(8 marks)*
*SPaG: 3 marks*

**Extra space**

..................................................................................................................................

..................................................................................................................................

..................................................................................................................................

..................................................................................................................................

..................................................................................................................................

**2** Tourist areas often become less popular over time. List **two** plans your chosen area has developed to cope with its problems and maintain a successful tourism industry for the future.

1 ...............................................................................................................................

..................................................................................................................................

2 ...............................................................................................................................

..................................................................................................................................

*(2 marks)*

Had a go ☐  Nearly there ☐  Nailed it! ☐

 Case study

# Mass tourism: good or bad?

Study this photograph of a safari in Kenya's Maasai Mara game park.

 tier F&H

**1** Use the photograph and your own knowledge to explain how mass tourism can have serious environmental effects.

....................................................................

....................................................................

....................................................................

....................................................................

....................................................................

....................................................................

....................................................................

....................................................................

....................................................................

....................................................................

> Try to link your reasons to evidence you have found in the photograph or from your own knowledge. It is always a good idea for questions like this to put in some relevant detail or a brief relevant example if you know one.

*(4 marks)*

**Case study**

# Keeping tourism successful

The following text extract describes part of the Kenyan National Tourism Policy (2006).

> Kenya's tourism industry has always been based on beach holidays and wildlife safaris. But Kenya can also offer a wide range of other types of holiday, such as golf, mountaineering, rock climbing, birdwatching, white water rafting, horse riding and camel treks, etc. Also Kenya has an extremely rich and interesting culture that many tourists would find fascinating.
>
> At the moment, most tourists visit quite a small number of places in Kenya: the beach resorts and the game parks. Developing cultural tourism would mean we could spread tourism to new areas, away from the most-visited and sometimes overcrowded areas. Developing cultural tourism would also help local people more because tourists like to buy handicraft products such as wood carvings, beadwork and paintings.

**tier Fn**

**1 (a)** What sort of holidays do most tourists go to Kenya for at the moment?

........................................................................................................................................

*(1 mark)*

**tier F&H**

**(b)** What is 'cultural tourism'?

........................................................................................................................................

*(1 mark)*

**tier F&H**

**> Guided >**

**(c)** Explain why the Kenyan government would want to expand the places that tourists visit in Kenya.

At the moment, most tourists have beach holidays or wildlife safari holidays in Kenya. This puts a lot of environmental pressure on a small number of places; it means areas get crowded and it means only a few areas in Kenya enjoy the economic benefits of tourism. So ...................................................................................

.................................................................................

.................................................................................

.................................................................................

| This answer has been started off for you. See if you can develop a couple more points that link to the explanation so far, to complete the answer. |

.................................................................................

.................................................................................

.................................................................................

.................................................................................

*(6 marks + 3 marks SPaG (F), 6 marks (H))*

**Extra space**

........................................................................................................................................

........................................................................................................................................

........................................................................................................................................

**101**

# Extreme tourism

1  Read this blog post from a tourist who has recently visited Antarctica.

> What a trip – truly the experience of a lifetime! Nothing I had read or seen prepared me for actually being there in the most beautiful, mesmerising and unspoiled landscape in the world. I'm so glad we went on a smaller ship because only 100 people are allowed to land at a time. Because there were only around 100 people on board, we got to go on shore every time. I would strongly advise anyone thinking of a trip to Antarctica to make sure the tour operators are signed up to the IAATO guidelines (Antarctic tour operators association), as this will guarantee your trip does not have a negative impact on the environment there: that has got to be the top priority here, people!

**(a)** The extract mentions one attempt to reduce the impact of tourism on Antarctica. What is it and how would it help reduce the damage done by tourists?

> When you use an extract like this to help you answer a question, try not to copy out big chunks from the extract – that just wastes your time. Instead, summarise the point or points it makes briefly, so you can concentrate on answering the question.

...........................................................................................................

...........................................................................................................

...........................................................................................................

...........................................................................................................

...........................................................................................................

...........................................................................................................

...........................................................................................................

...........................................................................................................

...........................................................................................................

...........................................................................................................

*(4 marks)*

**(b)** Outline how tourists can easily damage extreme environment areas.

...........................................................................................................

...........................................................................................................

...........................................................................................................

*(2 marks)*

**(c)** Describe **one** other measure that is used to reduce the impact of tourism in a named area with an extreme environment.

...........................................................................................................

...........................................................................................................

...........................................................................................................

*(2 marks)*

# Ecotourism

Study the following statements about ecotourism.

**1 (a)** Complete the table by ticking the correct box to show whether each statement is **True** or **False**.

| Statement | True | False |
|---|---|---|
| Ecotourism is organised tourism for large numbers of people going on holiday to the same place. | | |
| Ecotourism aims to help local people by creating jobs for them and helping them make money from what they sell. | | |
| Ecotourism can only happen in extreme environments where ecosystems are very fragile. | | |

*(3 marks)*

**(b)** Explain the **difference** between conservation and stewardship.

.................................................................................................................................

.................................................................................................................................

.................................................................................................................................

*(2 marks)*

**(c)** Give **two** reasons why tourists would be attracted to an ecotourism holiday.

1 ...............................................................................................................................

.................................................................................................................................

.................................................................................................................................

2 ...............................................................................................................................

.................................................................................................................................

.................................................................................................................................

*(2 marks)*

> If you run out of time while writing your last answers, you can just write down notes for answers. It won't be as good as a proper answer, but it will certainly be better than nothing.

# Stimulus materials –
# an introduction

Study this table showing changes in services in a part of Aberdeenshire in Northern Scotland.

| Services | 1981 | 2001 | 2011 |
|---|---|---|---|
| Shops | 254 | 217 | 110 |
| Primary schools | 36 | 32 | 22 |
| Post offices | 36 | 25 | 22 |
| Petrol stations | 34 | 23 | 23 |
| Doctors' surgeries | 10 | 5 | 0 |

**1 (a)** Complete this bar graph which plots the data for shops provided by the table above.

*(2 marks)*

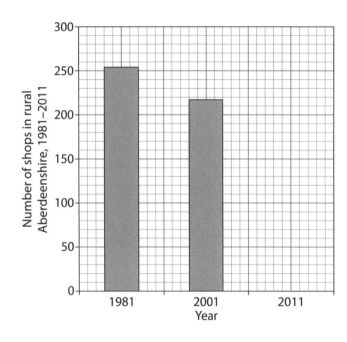

> Use a ruler when completing bar graphs. A ruler will also help you read off values accrately, which will help your answers.

**(b)** Describe the changes in services shown in the table above.

..................................................................................................................................

..................................................................................................................................

..................................................................................................................................

..................................................................................................................................

..................................................................................................................................

..................................................................................................................................

..................................................................................................................................

..................................................................................................................................

*(4 marks)*

> Make sure you use the figures to help you answer the question. Note the general trend and quote a few figures to illustrate it.

# Using and interpreting photos

Study this photograph of tourists visiting Seymour Island, Antarctica.

**1** Using evidence from the photo, describe:

- one reason why tourists want to visit Antarctica
- one way in which tourism could damage the Antarctic ecosystem
- one way in which tourism is controlled in Antarctica.

Reason for tourists visiting Antarctica

...................................................................................................................................................................

...................................................................................................................................................................

*(1 mark)*

How tourism could damage Antarctica ecosystems

...................................................................................................................................................................

...................................................................................................................................................................

*(1 mark)*

How tourism is controlled in Antarctica

...................................................................................................................................................................

...................................................................................................................................................................

*(1 mark)*

> The question asks for evidence from the photograph, so make sure you link what you say to something you can see or infer from the photo.

# Labelling and annotating

Study this photograph of the Steingletscher glacier in Switzerland in 2006.

**1** Draw a labelled sketch of the photograph to describe the evidence of recent glacial retreat.

*(4 marks)*

Draw a frame first to fit your sketch in, and then draw in the major features to give you fixed points to work to. Use as many labels as there are marks for the question and draw and label only what the photo shows you – don't invent any new features.

# Graph and diagram skills

Study this graph, which shows the growth in bicycle use observed (by traffic count) in New York City.

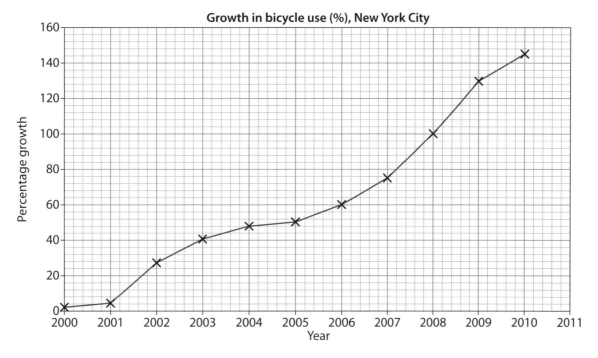

**1 (a)** Complete the graph by plotting the 2011 figure of 150 per cent.

*(1 mark)*

**(b)** What was the percentage growth in bicycle use observed between 2006 and 2009?

................................. per cent

*(1 mark)*

**(c)** How long did it take observed bicycle use to increase by 100 per cent?

...................................................................................................................................................

...................................................................................................................................................

*(1 mark)*

> Try not to rush questions like this – it is easy to make a silly mistake.

# Map types

Study this map, which shows the global distribution of countries according to levels of income, and the table below, which shows GNI per capita for selected countries.

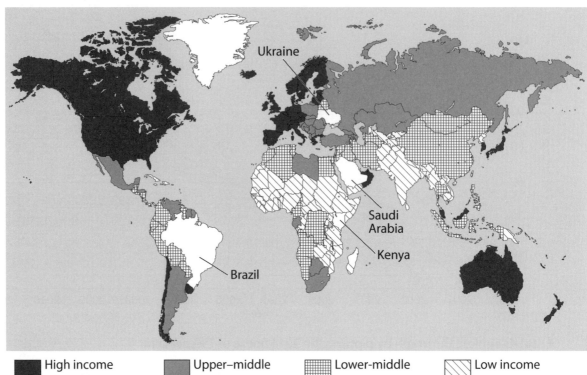

| High income countries US$11 116 or more | Upper–middle income countries US$3596–$11 115 | Lower-middle income countries US$906–$3595 | Low income countries US$905 or less |

| Country | GNI per capita (2011) |
| --- | --- |
| Brazil | $10 720 |
| Kenya | $820 |
| Saudi Arabia | $17 820 |
| Ukraine | $3120 |

1  (a)  Complete the map by shading in Brazil, Kenya, Saudi Arabia and Ukraine according to the key provided on the map.

*(4 marks)*

(b)  What type of map is shown above? Circle the correct answer.

**Isoline      Topological      Proportional symbol      Choropleth**

*(1 mark)*

> It is important to select the right shading for each country according to the key, but your shading in doesn't need to be perfect, just clear and accurate

# Describing maps

Study this map, which shows the global distribution of three ecosystems.

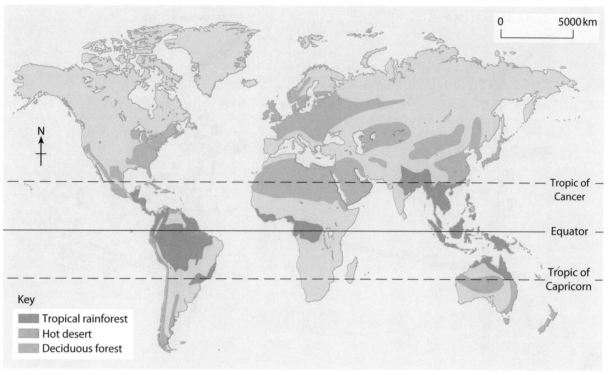

Key
- Tropical rainforest
- Hot desert
- Deciduous forest

**1** Describe the distribution of hot deserts shown in the map.

......................................................................................

......................................................................................

......................................................................................

......................................................................................

......................................................................................

......................................................................................

......................................................................................

......................................................................................

......................................................................................

......................................................................................

> Remember that if a question says 'describe' a distribution, it does not want you to explain the distribution. For a global distribution like this, you can use latitude, hemispheres and names of continents to structure your description, as well as compass directions.

*(4 marks)*

# Comparing maps

Study both maps, which show Mappleton, a village on the Holderness coast, in 1910 and in 1990.

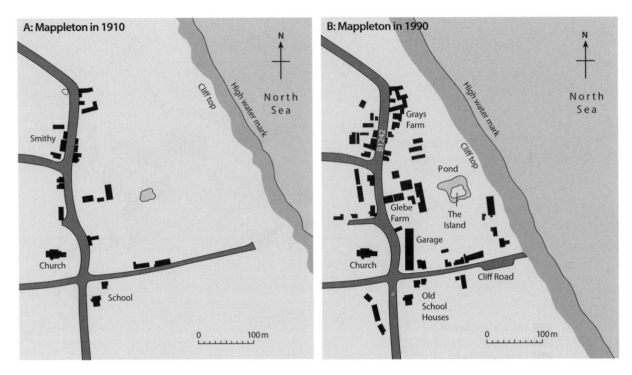

**1 (a)** By how many metres had the clifftop receded towards Mappleton between 1910 and 1990?

.................................................................................................................................................................

*(1 mark)*

**(b)** How far in metres was the straight line distance between the church and the high water mark in 1990?

.................................................................................................................................................................

*(1 mark)*

> Always, always, always check the scale of any map you are working with! Make sure any measurements you make are correctly converted to distances using the map's scale.

# Exam skills

1  Look carefully at the command terms in the table below that are often used in exam questions.
Connect each of the command terms on the left with their correct definition on the right.

| | |
|---|---|
| **Describe** | Make a list. |
| **Explain** | Come to a conclusion after giving different points of view. |
| **List** | Write down the most important points. |
| **Compare** | Write about the reasons for something; explain why. |
| **Contrast** | Add labels with details. |
| **Outline** | Write about the differences. |
| **Annotate** | Use a case study in your answer. |
| **To what extent...?** | Write about the similarities and differences. |
| **Use a named example...** | Write about what you can see. |

# Exam skills: using case studies

Use this checklist to make a list of all the case studies you need to revise. Remember, you will be doing:

- **3 topics** from Unit 1 – **one** from section A, **one** from section B and **one** from **either** A or B.

- **3 topics** from Unit 2 – **one** from section A, **one** from section B and **one** from **either** A or B.

> Make a revision flash card for each of your case studies, making a note of all the key information. You need to know the name and location plus the important points for each study such as: effects, responses, causes, advantages, disadvantages, uses, similarities, differences, challenges, management and so on.

| Tick your topics | Unit and topic | Case study requirements | My case studies – name and location |
|---|---|---|---|
| | **Unit 1 Section A The Restless Earth** | Fold mountain range | |
| | | Volcanic eruption | |
| | | Earthquake in a rich part of the world | |
| | | Earthquake in a poor part of the world | |
| | | Tsunami | |
| | **Unit 1 Section A Rocks, Resources and Scenery** | How people use granite landscapes | |
| | | How people use Carboniferous limestone landscapes | |
| | | How people use chalk and clay landscapes | |
| | | A working quarry | |
| | | Management of a quarry | |
| | **Unit 1 Section A Challenge of Weather and Climate** | Tropical storm in a poor part of the world | |
| | | Tropical storm in a rich part of the world | |
| | **Unit 1 Section A Living World** | Temperate deciduous woodland | |
| | | Tropical rainforest | |
| | | Hot desert – rich area of the world | |
| | | Hot desert – poor area of the world | |
| | **Unit 1 Section B Water on the Land** | Flooding in a rich part of the world | |
| | | Flooding in a poor part of the world | |
| | | Dam / reservoir in the UK | |

| Tick your topics | Unit and topic | Case study requirements | My case studies – name and location |
|---|---|---|---|
| | **Unit 1 Section B Ice on the Land** | Glacier case study | |
| | | Alpine area for winter sports / glacier sightseeing | |
| | **Unit 1 Section B The Coastal Zone** | Coastal flooding | |
| | | Recent or threatened cliff collapse | |
| | | Coastal management | |
| | | Coastal habitat | |
| | **Unit 2 Section A Population Change** | China's population policy since the 1990s | |
| | | Non birth-control population policy | |
| | | EU country with an ageing population | |
| | **Unit 2 Section A Changing Urban Environments** | Squatter settlement redevelopment | |
| | | Sustainable urban living | |
| | **Unit 2 Section A Changing Rural Environments** | Rural area in the UK | |
| | | Commercial farming area in the UK | |
| | **Unit 2 Section B The Development Gap** | A natural hazard | |
| | | A development project | |
| | | Contrasting EU countries: Wealthy country | |
| | | Contrasting EU countries: Poorer country | |
| | **Unit 2 Section B Globalisation** | One TNC | |
| | | Development of China as the new economic giant | |
| | | One type of renewable energy | |
| | **Unit 2 Section B Tourism** | UK National Park or coastal resort | |
| | | Tropical tourist area | |
| | | Extreme environment | |
| | | Ecotourism | |

# Unit 1 Physical Geography

The questions listed below will make up a practice paper. This will help you practise what you have learned, but may not be representative of a real exam paper. The pages on which these questions appear are listed below.

> Time: 1 hour 30 minutes
>
> You need to answer questions on three topics. Answer questions on **one** topic from **Section A** and one topic from **Section B** plus one from **either A** or **B**.

## Section A

Answer at least **one** question but **not more than two** questions in this section.

### 1. The Restless Earth

| | | |
|---|---|---|
| **1a)** | Page 1, question 1 | *(4 marks)* |
| **1b i)** | Page 2, question 1(a) | *(1 mark)* |
| **1b ii)** | Page 2, question 1(b) | *(3 marks)* |
| **1c i)** | Page 5, question 1(a) | *(2 marks)* |
| **1c ii)** | Page 5, question 1(b) | *(4 marks)* |
| **1d i)** | Page 7, question 1(a) | *(1 mark)* |
| **1d ii)** | Page 7, question 1(c) | *(4 marks)* |
| **1e)** | Page 9, question 1 | *(6 marks)* |

**Total for this section: 25 marks**

### 2. Rocks, Resources and Scenery

| | | |
|---|---|---|
| **2a i)** | Page 10, question 1a) | *(2 marks)* |
| **2b)** | Page 12, question 1 | *(6 marks)* |
| **2c)** | Page 13, question 1 | *(2 marks)* |
| **2d)** | Page 14, question 1 | *(3 marks)* |
| **2e)** | Page 14, question 2 | *(4 marks)* |
| **2f)** | Page 15, question 1 | *(4 marks)* |

**Total for this section: 25 marks**

### 3. Challenge of Weather and Climate

| | | |
|---|---|---|
| **3a i)** | Page 17, question 1(a) | *(2 marks)* |
| **3a ii)** | Page 17, question 1(b) | *(2 marks)* |
| **3b)** | Page 18, question 1 | *(4 marks)* |
| **3c)** | Page 21, question 1 | *(1 mark)* |
| **3d)** | Page 21, question 2 | *(2 marks)* |
| **3e i)** | Page 22, question 1(a) | *(2 marks)* |
| **3e ii)** | Page 22, question 1(b) | *(6 marks)* |
| **3f)** | Page 23, question 1 | *(2 marks)* |
| **3g)** | Page 23, question 2 | *(4 marks)* |

**Total for this section: 25 marks**

### 4. Living World

| | | |
|---|---|---|
| **4a)** | Page 26, question 1 | *(2 marks)* |
| **4b)** | Page 26, question 2 | *(3 marks)* |
| **4c)** | Page 29, question 1(b) | *(4 marks)* |
| **4d)** | Page 27, question 1(b) | *(6 marks)* |
| **4e i)** | Page 30, question 1(a) | *(4 marks)* |
| **4e ii)** | Page 30, question 1(b) | *(2 marks)* |
| **4f i)** | Page 31, question 1(a) | *(1 mark)* |
| **4f ii)** | Page 31, question 1(a) | *(3 marks)* |

**Total for this section: 25 marks**

## Section B

Answer at least **one** question but **not more than two** questions in this section.

### 5. Water on the Land

| | | |
|---|---|---|
| **5a i)** | Page 34, question 1(a) | *(4 marks)* |
| **5a ii)** | Page 34, question 1(b) | *(3 marks)* |
| **5b)** | Page 35, question 1 | *(4 marks)* |
| **5c i)** | Page 36, question 1(a) | *(3 marks)* |
| **5c ii)** | Page 36, question 1(b) | *(4 marks)* |
| **5d)** | Page 39, question 1 | *(1 mark)* |
| **5e)** | Page 39, question 2 | *(6 marks)* |

**Total for this section: 25 marks**

### 6. Ice on the Land

| | | |
|---|---|---|
| **6a i)** | Page 42, question 1(a) | *(1 mark)* |
| **6a ii)** | Page 42, question 1(b) | *(4 marks)* |
| **6b i)** | Page 43, question 1(a) | *(2 marks)* |
| **6b ii)** | Page 43, question 1(b) | *(1 mark)* |
| **6c i)** | Page 44, question 1(a) | *(3 marks)* |
| **6c ii)** | Page 44, question 1(b) | *(2 marks)* |
| **6d)** | Page 45, question 1 | *(3 marks)* |
| **6e)** | Page 45, question 2 | *(2 marks)* |
| **6f)** | Page 47, question 1 | *(1 mark)* |
| **6g)** | Page 49, question 1 | *(6 marks)* |

**Total for this section: 25 marks**

### 7. The Coastal Zone

| | | |
|---|---|---|
| **7a i)** | Page 50, question 1(a) | *(1 mark)* |
| **7a ii)** | Page 50, question 1(b) | *(4 marks)* |
| **7b)** | Page 53, question 1(a) | *(1 mark)* |
| **7c i)** | Page 56, question 1(a) | *(4 marks)* |
| **7c ii)** | Page 56, question 1(b) | *(4 marks)* |
| **7d i)** | Page 57, question 1(a) (i) | *(1 mark)* |
| **7d ii)** | Page 57, question 1(b) | *(2 marks)* |
| **7d iii)** | Page 57, question 1(c) | *(2 marks)* |
| **7e)** | Page 55, question 1(b) | *(6 marks)* |

**Total for this section: 25 marks**

# Unit 1 Physical Geography

The questions listed below will make up a practice paper. This will help you practise what you have learned, but may not be representative of a real exam paper. The pages on which these questions appear are listed below.

> Time: 1 hour 30 minutes
>
> You need to answer questions on three topics. Answer questions on **one** topic from **Section A** and one topic from **Section B** plus one from **either A** or **B**.

## Section A

Answer at least **one** question but **not more than two** questions in this section.

### 1. The Restless Earth

### 2. Rocks, Resources and Scenery

### 3. Challenge of Weather and Climate

### 4. Living World

## Section B

Answer at least **one** question but **not more than two** questions in this section. Use your case studies to support your answers where appropriate.

### 5. Water on the Land

### 6. Ice on the Land

### 7. The Coastal Zone

# Unit 2 Human Geography

The questions listed below will make up a practice paper. This will help you practise what you have learned, but may not be representative of a real exam paper. The pages on which these questions appear are listed below.

> Time: 1 hour 30 minutes
>
> You need to answer questions on three topics. Answer questions on **one** topic from **Section A** and one topic from **Section B** plus one from **either A** or **B**. You will be assessed on spelling, punctuation and grammar in this Unit.

## Section A

Answer at least **one** question but **not more than two** questions in this section.

### 1. Population Change
| | | |
|---|---|---|
| **1a)** | Page 59, question 1(b) | *(4 marks)* |
| **1b i)** | Page 60, question 1(b) | *(3 marks)* |
| **1b ii** | Page 60, question 1(c) | *(2 marks)* |
| **1c i)** | Page 61, question 1(a) | *(2 marks)* |
| **1c ii)** | Page 61, question 1(b) | *(2 marks)* |
| **1d)** | Page 62, question 1 | *(4 marks)* |
| **1e i)** | Page 65, question 1(b) | *(2 marks)* |
| **1e ii)** | Page 65, question 1(c) | *(6 marks) + SPaG: 3 marks* |

**Total for this section:** 25 marks + SPaG: 3 marks

### 2. Changing Urban Environments
| | | |
|---|---|---|
| **2a i)** | Page 67, question 1(a) | *(3 marks)* |
| **2a ii)** | Page 67, question 1(b) | *(2 marks)* |
| **2b i)** | Page 68, question 1 (a) | *(1 mark)* |
| **2b ii)** | Page 68, question 1 (b) | *(2 marks)* |
| **2c i)** | Page 69, question 1(a) | *(1 mark)* |
| **2c ii)** | Page 69, question 1(b) | *(6 marks) + SPaG: 3 marks* |
| **2d i)** | Page 71, question 1(a) | *(4 marks)* |
| **2d ii)** | Page 71, question 1(b) | *(4 marks)* |
| **2e)** | Page 73, question 1(b) | *(2 marks)* |

**Total for this section:** 25 marks + SPaG: 3 marks

### 3. Changing Rural Environments
| | | |
|---|---|---|
| **3a i)** | Page 75, question 1(a) | *(3 marks)* |
| **3a ii)** | Page 75, question 1(b) | *(1 mark)* |
| **3b i)** | Page 76, question 1(a) | *(2 marks)* |
| **3b ii)** | Page 76, question 1(b) | *(4 marks)* |
| **3c)** | Page 77, question 1(b) | *(4 marks)* |
| **3d i)** | Page 80, question 1(a) | *(1 mark)* |
| **3d ii)** | Page 80, question 1(b) | *(6 marks) + SPaG: 3 marks* |
| **3e)** | Page 81, question 1(b) | (4 marks) |

**Total for this section:** 25 marks + SPaG: 3 marks

### 4. The Development Gap
| | | |
|---|---|---|
| **4a i)** | Page 82, question 1(a) | *(1 mark)* |
| **4a ii)** | Page 82, question 1(b) | *(2 marks)* |
| **4b)** | Page 83, question 1(a) | *(3 marks)* |
| **4c)** | Page 84, question 1 | *(6 marks) + SPaG: 3 marks* |
| **4d i)** | Page 85, question 1(a) | *(2 marks)* |
| **4d ii)** | Page 85, question 1(b) | *(2 marks)* |
| **4e)** | Page 86, question 1(a) | *(1 mark)* |
| **4f i)** | Page 87, question 1(a) | *(4 marks)* |
| **4f ii)** | Page 87, question 1(b) | *(4 marks)* |

**Total for this section:** 25 marks + SPaG: 3 marks

## Section B

Answer at least one question but **not more than two** questions in this section.

### 5. Globalisation
| | | |
|---|---|---|
| **5a i)** | Page 89, question 1(a) | *(3 marks)* |
| **5a ii)** | Page 89, question 1(b) | *(1 mark)* |
| **5b)** | Page 91, question 1 | *(4 marks)* |
| **5c)** | Page 92, question 1 | *(6 marks) + SPaG: 3 marks* |
| **5d)** | Page 93, question 1 | *(4 marks)* |
| **5e)** | Page 93, question 2 | *(1 mark)* |
| **5f i)** | Page 94, question 1(a) | *(1 mark)* |
| **5f ii)** | Page 94, question 1(b) | *(4 marks)* |
| **5f iii)** | Page 94, question 1(c) | *(1 mark)* |

**Total for this section:** 25 marks + SPaG: 3 marks

### 6. Tourism
| | | |
|---|---|---|
| **6a i)** | Page 96, question 1(a) | *(1 mark)* |
| **6a ii)** | Page 96, question 1(b) | *(1 mark)* |
| **6b)** | Page 96, question 2 | *(3 marks)* |
| **6c i)** | Page 97, question 1(a) | *(1 mark)* |
| **6c ii)** | Page 97, question 1(b) | *(1 mark)* |
| **6d)** | Page 97, question 2 | *(4 marks)* |
| **6e i)** | Page 98, question 1(a) | *(1 mark)* |
| **6e ii)** | Page 98, question 1(b) | *(3 marks)* |
| **6f)** | Page 99, question 2 | *(2 marks)* |
| **6g i)** | Page 101, question 1(a) | *(1 mark)* |
| **6g ii)** | Page 101, question 1(b) | *(1 mark)* |
| **6g ii)** | Page 101, question 1(c) | *(6 marks) + SPaG: 3 marks* |

**Total for this section:** 25 marks + SPaG: 3 marks

# Unit 2 Human Geography

The questions listed below will make up a practice paper. This will help you practise what you have learned, but may not be representative of a real exam paper. The pages on which these questions appear are listed below.

> Time: 1 hour 30 minutes
>
> You need to answer questions on three topics. Answer questions on **one** topic from **Section A** and one topic from **Section B** plus one from **either A** or **B**. You will be assessed on spelling, punctuation and grammar in this Unit.

## Section A

Answer at least **one** question but **not more than two** questions in this section.

### 1. Population Change

### 2. Changing Urban Environments

### 3. Changing Rural Environments

### 4. The Development Gap

## Section B

Answer at least **one** question but **not more than two** questions in this section.

### 5. Globalisation

### 6. Tourism

# Answers

## UNIT 1

### The Restless Earth

#### 1. Unstable crust
1  True, false and true.

#### 2. Plate margins
1  **(a)**  A destructive plate margin.
   **(b)**  **A** = a volcano; **B** = an oceanic trench;
   **C** = a subduction zone.

#### 3. Fold mountains and ocean trenches
1  **(a)**  A destructive plate margin.
   **(b)**  A geosyncline is formed. This is filled with sediment which is compressed over millions of years into sedimentary rock. Plate movements fold the rock layers into fold mountains.

#### 4. Fold mountains
1  Named example: Alps, Andes. Answers will depend on the chosen case study area. A really good answer would talk about two key challenges of the area, such as there being only small areas of flat land to farm on, or the difficulties of communicating between valleys. These should be linked to the ways people have traditionally adapted to them or have used modern technology or opportunities to get round them.

#### 5. Volcanoes
1  **(a)**  A, B, C.
   **(b)**  Differences could include:
   • the size of the eruption (an eruption of at least 1000 km³ of magma)
   • the global effect of a supervolcano eruption (likely to lead to a global volcanic winter due to all the dust circulating in the atmosphere)
   • the way they are formed – a massive magma chamber underlying the surface, pushing it up into a dome
   • supervolcanoes also form a caldera when they erupt.

#### 6. Volcanoes as hazards
1  Reasons may include two of the following:
   • some eruptions happen in populated areas and some in unpopulated areas
   • some eruptions happen in richer countries that can afford to monitor volcanoes for eruptions and so can evacuate people and defend property
   • composite volcanoes produce violent and more dangerous eruptions that are infrequent and irregular; shield volcanoes produce gentler eruptions that are frequent and regular so people are expecting them and can make sure they stay safe
   • nature of the lava and its gas content.

#### 7. Earthquakes
1  **(a)**  Maule region, Chile.
   **(b)**  Haiti.
   **(c)**  Two from: could have occurred in an area of low population density, in an area with earthquake-proof buildings, in an area where the population was well prepared for earthquakes and knew what to do to improve their survival chances, in an area with really good immediate responses to the earthquake; physical reasons could be that the earthquake had its focus deep underground so the shockwaves were weaker at the surface, the earthquake could have been in hard rock rather than sands and clays.

#### 8. Earthquake hazards
1  A second point could be about:
   • the characteristics of the earthquake itself: magnitude can differ, whether it is near the surface or deep underground, whether the focus is in solid rock or sands and clays which vibrate more strongly
   • the nature of the human population: in some earthquake-hit areas there are very few people, living at very low densities, and an earthquake there will have a very different impact than on a high-density city, packed with people
   • the time of day: whether people are at home asleep or travelling to work, or packed into offices and schools.

#### 9. Tsunamis
1  This answer will depend on the case study you are using. You do need to give the name of the example you are using, for example the Asian tsunami of December 2004 or the Japan earthquake of March 2011, or your answer will be very limited in how well it will do. Good answers will describe the effects very clearly and the answer will be well organised. A good answer will also link your statements together, for example: 'Because there were very large numbers of people killed, there were many dead bodies in the rubble. This led to the risk of diseases spreading among the survivors'.

### Rocks, Resources and Scenery

#### 10. Rock groups
1  **(a)**  Y is granite and Z is chalk.
   **(b)**  Chalk, clay or limestone could all be the answer given for a sedimentary rock; granite for the igneous rock.

#### 11. The rock cycle
1  **(a)**  Chemical weathering.
   **(b)**  Your answer should have two points or one developed point. The first should describe how rainwater combines with carbon dioxide in the atmosphere to form carbonic acid. The second should explain how carbonic acid dissolves calcium carbonate by changing it into calcium bicarbonate which is soluble and so is removed by water.

#### 12. Granite landscapes
1  The description could also include points about the colour of the rock, the lack of vegetation cover on the rock, what the landscape around the tor looks like. For the formation part of the answer, your answer should explain how there is more weathering of granite that has lots of joints close together than those that are far apart. You should refer to the type of weathering: chemical weathering when the tor was under the surface, greater in the areas that were more jointed, and then when the surface weathered rock was removed during the Ice Age, the tor was exposed to freeze-thaw weathering that reduced it down to its current form.

#### 13. Carboniferous limestone landscapes
1  The correct sequence is: Cave = **Y**, Gorge = **Z**, Limestone pavement = **W**, Resurgence stream = **X**, Swallow hole = **V**.

#### 14. Chalk and clay landscapes
1  The chalk escarpment is the higher land, the clay vale is the flat valley and the scarp slope is the steep slope forming the edge of the escarpment, dipping down to the clay vale.
2  The completed sketch should look something like this:

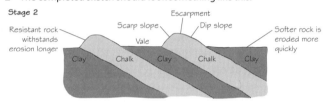

#### 15. Quarrying
1  A good answer will develop the statements you make about the map extract and will also make links between the points. For example, the map shows a railway from the works that skirts the village of Hope. You should then make a link between this to the environmental disadvantages of rail trucks transporting cement and stone: could be noisy and possibly dirty for the residents of Hope.

#### 16. Quarrying management strategies
1  Your answer needs to consider both parts of the question: strategies used while the quarry was in operation and strategies used as the quarry was worked out, or some parts of it were no longer worked. The question wants a named example such as Hope Quarry, so you must provide this or it will seriously damage how good your answer can be. In your use of case study information, you should provide details and examples to support your answer that are relevant to the question. Your statements need to have detailed development, so you should not just list points but you should develop them and make links between them.

### Challenge of Weather and Climate

#### 17. The UK climate
1  **(a)**  The words in the right order are: north, west, east.
   **(b)**  Your answer could make two points, such as: the prevailing wind from the south-west brings moisture-rich air to the west (Aberystwyth); relief rainfall means by the time the air reaches Norwich it has lost most of its moisture. Or you could make one developed point.

#### 18. Depressions
1  **A** = Depression, **B** = Warm front, **C** = Cold front, **D** = Occluded front, **E** = Isobar and **F** = Anticyclone.

#### 19. Anticyclones
1  **(a)**  An anticyclone.
   **(b)**  Answers might include: hot temperatures because there are no clouds to block the Sun; little wind or light winds only because of the gentle pressure gradient of the high pressure system; no rain because the warm air can hold more moisture.
   **(c)**  A blocking anticyclone sits in the path of low pressure systems coming in from the Atlantic and diverts them around it.

#### 20. Extreme weather in the UK
1  Impacts could include:
   • health (physical health issues and mental health issues like depression from being unable to return to their flood-damaged homes until they are restored)
   • economic impacts, for example agriculture being affected
   • transport impacts – roads being closed, bridges being damaged, etc
   • costs of making repairs and of accommodating people until the repairs are made. Increased insurance premiums.

**21. Global warming**

**1**  B

**2**  Choose from the list below.
- There has been a change in around 1 °C over the period.
- The general trend has been a rising one with fluctuations.
- There has been a trend towards more rapid temperature increase in the period since the late 1950s.

**22. Consequences of global climate change**

**1  (a)**  Any one of: a decline in food production; increased tropical storm activity; sea level rise leading to increased flooding; secondary causes of increased conflict.

**(b)**  Development of first point: this could lead to hunger, malnutrition and in worse cases starvation. It adds increased stress to an already poor population, especially if yields drop by more than the estimated 30% per cent.
Choose a second point to develop, for example: cyclones are a major threat to human life, property and communications networks. They can cause flooding and water shortages. People may become ill or die from water-borne diseases. Many people may be made homeless when buildings are destroyed. Roads and railways may be destroyed, making communication very difficult. Try to make links between your points, refer to information provided by the source, make use of your own knowledge if relevant and keep your answer focused on the question. There is no credit for merely repeating the answers to part **(a)** without developing them.

**23. Responses to the climate change threat**

**1**  Two from: reducing waste, increasing recycling, using less paper and water, helping local homes and businesses become more energy efficient, reducing and controlling any pollution caused by council activities.

**2**  To develop the first point make a link between increased rainfall intensity, the increased run-off in urban centres and deforested slopes and the risk of flooding.
Two further developed points: drought management from reduced rainfall (especially low intensity winter rainfall); biodiversity loss links to increased pressures on wildlife and plant life caused by climate change (changing ecosystems); health risks link to the appearance or increase in severity of disease linked to warmer temperatures (such as malaria) and it includes diseases for humans but also for animals (livestock, pets) and plants (crop diseases).

**24. Causes of tropical revolving storms**

**1**  Include another two factors in your answer: the effect of latitude (between 5° and 15° north and south of the equator) due to the Coriolis effect, and low atmospheric shear; the Coriolis effect is the effect of the Earth's spin and low atmospheric shear describes a situation when wind speeds are the same all the way up through the atmosphere.
You might also bring in severe air instability. You may mention that no one is completely sure what makes a hurricane form but the distribution certainly suggests these factors are very important.

**25. Comparing tropical revolving storms**

**1**  This answer will depend on your case studies, e.g. Katrina, but, a really good answer will use precise facts and details of the economic, social or environmental impacts. The answer asks for one out of the three categories of effect, so you should make sure you have done that rather than covering all three.

## Living World

**26. What is an ecosystem?**

**2  (a)**  Example of a producer: rushes, water lily, green algae, some of the bacteria.

**(b)**  Example of a consumer: the frog, small fish, large fish, some bacteria, heron.

**(c)**  Example of decomposer: bacteria.

**27. Ecosystems**

**1  (a)**  30° N and S

**(b)**  Your answer should link, for example:
- storing water with enabling the plant to use its stores when there is no other water available
- or for plants that grow and flower very quickly after rain to produce a tough seed that lies dormant until the next time it rains in the desert
- or how some plants like the Joshua tree have adapted to reduce water loss from the high transpiration rates with features such as thick bark and spiky leaves
- or plants have very extensive and deep root systems to extract whatever water does become available in the desert soils.

**28. Temperate deciduous woodlands**

**1**  Answers might include: leaving dead wood to decay is sustainable because it ensures that nutrients in the dead wood are returned to the ecosystem rather than being lost and it also supports a lot of insect life, which in turn means more food for insect-eating creatures such as wood mice, badgers and hedgehogs.

**29. Deforestation**

**1  (a)**  Negative impacts are most acute for native peoples wishing to continue a traditional way of life in the rainforest:
- the pollution of local rivers and lakes caused by soil erosion

following deforestation
- the rapid loss of soil fertility that often accompanies the leaching and soil erosion that follows deforestation
- the introduction of diseases from outside the region can be fatal for indigenous people.

**(b)**  The question asks for two out of three causes deforestation, so you should make sure have you done that rather than covering all three.
- Commercial farming and ranching: large-scale farming needs the land to be cleared of almost all trees; it cannot be conducted as economically when crops or livestock are mixed in with tropical rainforest vegetation. There is a high demand for more food crops, for meat from livestock and more biofuels around the world, and this demand makes it very profitable to clear rainforest for farming. Commercial farming often follows logging.
- Mining is associated with tropical rainforest deforestation particularly in the case of open-cast mining: usually the cheapest way to get to buried minerals is to dig a massive pit rather than to dig mine shafts and mine the minerals from underground. Mining also needs transport out of the area: roads and railways, which also need forest to be cleared.
- While loggers are usually only interested in particular types of trees, it is much cheaper to clear the whole area of all trees and sell the land to commercial farmers than it is to select only a few trees for felling and extract those from the rainforest (selective logging). Logging is big business because many tropical trees produce excellent timber which can be sold for high prices.

**30. Sustainable management of tropical rainforest**

**1  (a)**  True (guided answer), false (it wants half that), true, false.

**(b)**  Selective logging, replanting, ecotourism, reducing demand for tropical hardwoods.

**31. Rainforest management**

**1  (a)**  The rate of deforestation has declined almost every year between 2004 and 2011, although it did increase very slightly from 2007 to 2008.

**(b)**  Answers might include three from:
- international agreements to reduce rainforest deforestation leading to national commitments (like Brazil's commitment to reduce deforestation to 1900 square miles a year by 2017)
- surveillance of illegal deforestation by satellite and then arrests and fines for those responsible
- protection of areas of the rainforest through national parks or other nature reserves
- debt for conservation swaps, in which poorer countries promise to protect areas of the forest in return for a reduction in international debts
- provision of alternative sources of income from the forest such as ecotourism or sustainable timber production rather than illegal logging; selective logging rather than wholesale logging
- improvement of cattle ranch soils to prevent new areas of pasture being required to replace old and worn-out ones every few years.

**32. Economic opportunities in hot deserts 1**

**1**  Your answer will depend on the case study you have chosen, e.g. Las Vegas. A good answer will refer to more than one type of use, will use relevant detail and examples from your case study and will make linkages between points and their development: for example, the hot desert areas have a hot, dry climate which is very popular as a tourism destination for older people who may be looking to escape from the cold winters where they live.

**33. Economic opportunities in hot deserts 2**

**1**  Your answer will depend on the case study examples you have chosen to use.
*Challenges:*
The common issue of access to water, living with very hot daytime temperatures and cold night temperatures.
*Differences:*
Nomadic pastoralism as the traditional poor world method of making use of meagre desert vegetation; mining, commercial farming and tourism in richer countries.

## Water on the Land

**34. Changes in the river valley**

**1  (a)**  **W** = Source, **X** = Watershed, **Y** = Tributary, **Z** = Mouth.

**(b)**  Upland, long profile, cross profile, flat.

**35. Erosion and transportation**

**1**  **W** = suspension, **X** = traction, **Y** = solution, **Z** = saltation.

**2  (b)**  Erosion is associated with a high rate of discharge as the increased volume or flow shows the river will have lots of surplus energy. Deposition will occur when discharge decreases and the river no longer has sufficient energy to transport as much of its load.

**36. Waterfalls and gorges**

**1  (a)**  Your descriptions should include three from: the hard, resistant rock 'cap'; the overhang, the less resistant rocks below the cap, the waterfall itself.

(b) Your completed diagram should resemble the diagram in the accompanying Revision Guide, without the information on how the waterfall retreats. You should use annotations, which describe how the waterfall is formed, not just labelled features on your diagram(s), e.g. undercutting of less resistant rock; collapse of resistant layer; downward erosion causing plunge pool; retreat upstream of waterfall.

## 37. Erosion and deposition

1 The first of these is shown in the guided answer; the river cliff label should be to the outside bend (probably up at the top of the picture is best); the slip-off slope is on the inside of the bend (shown quite nicely on the inside bend at the bottom of the picture); the floodplain is not very evident but any label inside the meander would be valid.

2 Your diagram should look something like the diagram on ox-bow lake formation in the accompanying Revision Guide. Make sure you have made the role of deposition and erosion really clear in your answer, and marked the direction of the river flow.

## 38. Flooding 1

1 (a) The rising limb should label the rising up part of the line graph, the falling limb the falling down part of the line graph and the lag time should show the distance in time between the peak of the precipitation and the peak of the discharge.

(b) River X is more likely to flood (it has a 'flashy' hydrograph).

(c) To continue the guided answer, you should have said: this means more surface run-off would have reached the river more quickly than Y, where more vegetation (forest) would have intercepted the water: some would have evaporated from the leaves, some would have been taken up by the trees and less run-off would reach the river and would reach it more slowly.

## 39. Flooding 2

1 Statement C.

2 To complete the guided part of the answer: the differences are due to richer countries having more money and so are more likely to have developed flood protection measures and to have included effective planning for flood events and evacuation of at-risk areas, which means responses to flood events can happen in a rapid fashion. Other reasons could be communications and infrastructure: it can often be difficult to reach affected parts of poorer countries if they are remote, while in richer parts of the world there are few areas that are difficult to access and emergency services and the military have access to helicopters to aid in evacuation. This may not be as easy to achieve in poorer countries, where lack of money to develop defences, monitoring and planning can hamper the speed and effectiveness of the response.

## 40. Hard and soft engineering

1 The answers are: building flood walls is hard engineering; planting hedges to divide up the car park is soft engineering; widening and deepening the river is hard engineering; moving the car park to a higher location is soft engineering.

2 This answer will depend on your selection of either hard or soft engineering. You may have decided to stress the advantages of your chosen option or the disadvantages of the option you did not choose – or a bit of both. All of these are valid ways to approach this question.

## 41. Managing water supply

1 The impact of taking lots of water out of a river would include making life more difficult for fish and many species of insects, birds and animals. There are also other human activities that depend on abstraction of river water, under licence from the local water agency, especially farming, power generation and some industries.

2 This answer will depend on your case study, e.g. Cow Green Reservoir, County Durham. Remember these issues can be positive as well as negative, for example the environmental impact of building a reservoir will mean the destruction of natural habitats, flooded under the reservoir, but then the reservoir can then become a valuable natural habitat of a different kind – many reservoirs have active and important conservation areas. Really good responses to this question would show good organisation of your points and examples and would not have many spelling, punctuation or grammar mistakes.

## Ice on the Land

### 42. Changes in ice cover

1 (a) The relationship is that higher levels of atmospheric $CO_2$ are associated with interglacials and lower levels of atmospheric $CO_2$ are associated with glacials.

(b) The rest of the answer should go on to explain how an ice core taken from a thick ice sheet provides a sample of all the different layers that make up the ice sheet and each of these layers can be analysed to identify the concentration of $CO_2$ in the atmospheric gas bubbles, and that recording tied to how long ago each layer was created.

### 43. The glacial budget

1 (a) Examples include: measurements of glacier length, old maps, aerial photos and satellite images, measurements from scientists studying glaciers over the years (some glaciers in Switzerland have been studied and measured for well over 100 years), monitoring data from GPS equipment on glaciers, landforms, debris and other

physical indications of a glacier's presence and retreat back up the valley.

(b) A negative glacial budget is when ablation exceeds accumulation.

### 44. Glacial weathering, erosion, transportation and deposition

1 (a) Transportation is shown by the moraine in the middle of the glacier; deposition is happening along its sides and also at the bottom right of the glacier, where a small branch of the glacier has split away.

(b) Your diagram should look something like this:

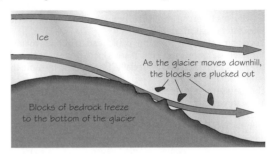

Ice

As the glacier moves downhill, the blocks are plucked out

Blocks of bedrock freeze to the bottom of the glacier

(c) Bulldozing is when a glacier has retreated, leaving a pile of moraine where its snout has melted, and then advanced again, pushing the moraine away in front of it.

### 45. Glacial erosion landforms 1

1 Compacts, weathering, erode, slip.

2 Two from: contours close together, contours curved round in a bowl shape, possibly a tarn, possibly a place name: cwm in Wales, cirque in the Alps (both mean corrie).

3 Your diagram will probably look something like this:

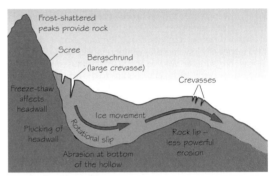

Frost-shattered peaks provide rock

Scree

Bergschrund (large crevasse)

Crevasses

Freeze-thaw affects headwall

Ice movement

Plucking of headwall

Rotational slip

Abrasion at bottom of the hollow

Rock lip – less powerful erosion

### 46. Glacial erosion landforms 2

1 There are several features indicating that this is a glaciated valley.

- The valley has very steep sides and a wide, flat bottom: the classic 'U shape' of the glacial trough. This indicates glaciation because a glacier is a very powerful eroding and transporting force that removes interlocking spurs of a river valley and deepens and broadens the valley as it passes through it.
- The waterfalls at Dungeon Ghyll Force and above it suggest a hanging valley; this valley probably had a smaller glacier in it that was formed in the corrie now occupied by Stickle Tarn; this didn't erode as deeply as the main glacier in the Langdale Valley and so was left hanging higher up than the main valley floor.
- Misfit stream – the 'Great Langdale Beck' looks much too small to be responsible for this deep, broad valley – this is also strong evidence of glaciation.

### 47. Glacial landforms of transportation and deposition

1 X is a pair of lateral moraines.

2 The landforms at Y look like an older pair of lateral moraines formed when the glacier was wider and bigger. The glacier must have retreated since they were formed (by freeze-thaw weathered rock and landslide debris falling down the valley sides) and then moved back up the valley as a smaller glacier, still carrying debris from the narrower valley sides further up the valley, to deposit a new set of lateral moraines inside the old ones. The glacier is now retreating again, leaving both sets of lateral moraines behind.

### 48. Alpine tourism – attractions and impacts

1 Your answer will depend on your Alpine area. Try to link your statements together, so for example: for environmental impacts, you might say that developers clear forested land in order to build more hotels to deal with rising numbers of tourists; for example, 100 square kilometres of forest have been cleared in the Alps for this reason. This forest clearance also has another environmental impact: it makes slopes less stable and so increases the risk of avalanches.

### 49. Management of tourism and the impact of glacial retreat

1 It is fine for you to question how far management can be sustainable, for example in the context of glacial retreat, but good answers will give examples of, such as: environmentally friendly public transport systems for tourists which solve the congestion problems and pollution caused by private cars, or the fencing off and reseeding of bare patches in grass cover caused by people skiing on slopes where the snow is too thin.

## The Coastal Zone

### 50. Waves and coastal erosion

1 **(a)** Mass movement
  **(b)** Reasons might include two from: water and a rock type susceptible to mass movement; removal of the natural protection provided by a sandy beach because longshore drift has been stopped by the building of groynes further up the coast; saturation of the rocks with water makes landslides more common (the rock is heavier and looser, and mass movement is lubricated by the water); the soft sandstone rock, which may have become softer and more weathered by the heavy rain; undercutting of the cliff by wave action, which would have contributed by making the slope less stable; people walking on top of the cliff.

### 51. Coastal transportation and deposition

1 **(a)** Your finished longshore drift diagram should look something like this:

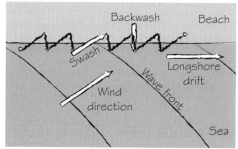

  **(b)** Your finished constructive wave diagram should look something like this:

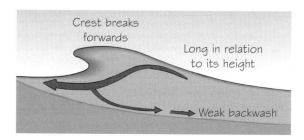

### 52. Landforms of coastal erosion

1 **(a)** **U** = Stump, **V** = Stack, **W** = Arch, **X** = Wave-cut platform, **Y** = Wave-cut notch, **Z** = Cliff.
  **(b)** Your finished diagram should look something like this:

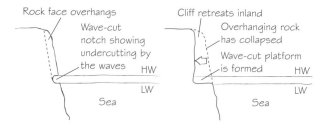

### 53. Landforms resulting from deposition

1 **(a)** A spit
  **(b)** Your completed diagram should look something like this:

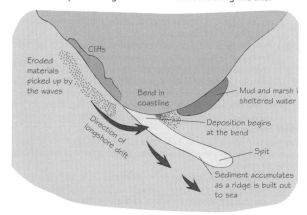

### 54. Rising sea levels

1 The details for this answer will depend on your case study area but good answers will have a clear structure which is likely to be divided into economic, social, political and environmental consequences.

Really good answers will show detailed understanding, which should be supported by relevant evidence and examples. And above all, if the question asks for a case study, make sure you use one and make sure you say what and where it is and that you include details that are specific to your chosen example.

### 55. Coastal management

1 Any one of the hard engineering strategies could be used here, for example a sea wall or rock armour for the exposed coastline areas, plus groynes for the more sheltered bay area where erosion from longshore drift might also be a problem.

2 Your answer might include some of the following points:
  *Costs:*
  • none if the coast is stable
  • loss of farmland
  • people losing their homes and businesses
  • loss of roads and pathways
  • social and political problems associated with people feeling they have been abandoned when they should be protected from this type of threat
  • environmental problems to do with loss of natural habitat.
  *Benefits:*
  • If the coast is stable then the money can be used elsewhere.
  • The coast may be at threat but not economically valuable enough to save. Money that may have been spent here may be better used in an area that has more need or is more valuable.
  • A do-nothing approach might simply mean that the coast along these stretches is already well protected either by natural or man-made defences and therefore nothing further needs to be done. In this case, the benefits of the scheme are that time and money do not need to be spent defending these areas and they can instead be focused on the areas that do need defending or managing, and there are no real costs, as long as the initial appraisal of the state of the coastal defences has been done properly. In other situations, however, the coast may be under very serious threat of erosion or flooding but it is not considered economically valuable enough to justify defending it. In this sort of case, the benefit remains that other areas that do have more reasons to be defended can have attention and money focused on them, but there are then a wide range of costs that you can explore in your answer, as listed under Costs above.

### 56. Cliff collapse

1 **(a)** Other factors could include: the composition and structure of the rocks (for example, if a permeable layer overlays an impermeable one, allowing the top layer to become saturated with water); the strength of the waves hitting the coastline (which will depend on the fetch of the winds – how much open sea prevailing winds blow across); whether the coastline is exposed to the sea or sheltered in bays; whether the cliff line has a wide beach or a narrow one in front of it (a wide beach can soak up a lot of wave energy); whether the coastline is exposed to higher than average sea level rises or a high tidal range; whether wave types are destructive or constructive.
  **(b)** There are a range of impacts: loss of life or accident (it doesn't look as if safety measures to protect people from falling over the cliff are very good in the photo); economic losses if the holiday homes are damaged by a cliff collapse; difficulties in getting affordable insurance for such properties; difficulties in attracting tourists to stay in a park or resort area with these types of problems.

### 57. Managing the coast

1 **(a)** **(i)** Rock armour
      **(ii)** Groynes
  **(b)** By trapping sediment that is being moved along the coast by longshore drift.
  **(c)** Beach nourishment has to be repeated fairly often – once a year in some cases – which makes it more expensive over the long term. It also increases deposition further down the coast.

### 58. Coastal habitats

1 This will depend on your chosen coastal habitat. Make sure you give as much relevant detail as you can to bring your named example to life. Link your strategies to the conservation needs of the area, for example: 'The Essex marshes have long been under threat as sea walls have been built and drainage systems developed to turn the marshland into farmland. In 2006, a trial scheme to extend the salt marsh habitat saw a sea wall being breached to allow the sea to flood an area to create more salt marsh habitat, which ensured more rare plant and animal species could be protected.' As well as conservation of the ecosystem, consider ways in which the habitat is being protected from damage by visitors, and how conflicts over land use are being resolved, for example creating boardwalks for visitors and making sure farmers are compensated for loss of land and have the benefits of salt marsh as a defence against storm damage explained to them.

# UNIT 2

## Population Change

### 59. Population explosion

1  (a)  There are two main trends: a fall in the death rate, including the infant mortality rate once data for this becomes available, and a fall in the birth rate. The fall in the death rate happens sooner and is more rapid. It becomes especially rapid when infant mortality rates are halved between 1950 and 2000. The fall in the birth rate happens later, only really getting going between 1950 and 2000. This staggered pattern would result in exponential population growth. By 2050, the prediction for the world's population is a very low birth rate and an only slightly lower death rate, leading to slow rates of natural increase.

   (b)  The death rate fell because of medical advances such as better medicines and the introduction of vaccination and immunisation schemes. More doctors and nurses meant medical attention is better. Houses have become more hygienic with indoor toilets and there is better sewage disposal and safer drinking water. People also generally have a better diet.

### 60. Demographic transition

1  (a)  Demographic Transition Model.
   (b)  Your complete key should look like this:

   ——— Birth rate
   ——— Death rate
   ——— Total population

   (c)  Brazil and India.

### 61. Population structure

1  (a)  It has a broad base. It also has steeply sloping concave sides.
   (b)  It has a declining birth rate shown by the narrowing base, a large bulge of middle-aged population, a large elderly population and a relatively high number of people surviving into extreme old age.

### 62. Growing pains

1  •  Emancipation of women: when women have more control over their lives this includes control over their own bodies and how often they have babies, whether or not they choose to use birth control or abortion services. When women have voting rights, woman politicians are more likely to be elected to political power and attitudes towards women's roles in society start to change. Women's health becomes more of a priority. Support increases for women being educated and going to work as a valued alternative to women staying at home and caring for children.

   •  Education: education increases the number of people who are able to get better paying jobs than subsistence agriculture, which relies on large families for manpower. It increases understanding of contraception and can decrease infant mortality because families are better able to safeguard their children's health through things like washing hands to reduce the spread of germs.

   •  Agricultural change: subsistence agriculture depends on large families to provide the labour for farming, and as subsistence agriculture does not give high yields, very nutritious food or very reliable crops then it is associated with high infant mortality. Any change towards higher yields, more reliable crops or, especially, increased income from farming will help reduce birth rates. However, this sort of agricultural change often results in large numbers of people having to leave farming as it becomes more mechanised and richer famers buy up all the land, which results in high rates of rural–urban migration and urbanisation.

   •  Increasing urbanisation lowers the birth rate because people no longer need big families to work the land, and people generally live better lives in the cities than in the countryside even if they are living in squatter settlements and working in the informal sector.

### 63. Managing population growth

1  This question wants you to talk about changes in the policy linked to the reasons for them, so be careful not to write about how the policy began or the ways it was enforced and encouraged in China. You can use the extract to help you remember changes that were linked to problems such as: an ageing population and the extra pressure that would put on families where both parents had no brothers and sisters; the recognition of the importance of maintaining the family name in China, which is what lies behind the preference for boys (this is linked to the relaxation in rural areas since a second child here is only allowed if the first child was a girl, or the first boy child is over 5 years old). You could also say that the policy is restricted to the Han Chinese, and not to the ethnic minorities because they are small in number, live in remote areas and do not contribute much to the economic growth of the country. Reference could also be made to parents in dangerous occupations being allowed to have more than one child, and also that families with a disabled child can have another child. Again, explain the reasons for these changes as well as stating them.

### 64. Ageing population

1  Your answer should talk about the declining birth rate as a second factor, developed to discuss the implications of fewer births meaning a smaller working population and so reduced tax reserves in the future to provide for a growing elderly dependent population.

### 65. Migration: push and pull

1  (a)  Refugees are migrants who have been forced to flee an area.
   (b)  There are lots of possible push factors: war, civil war (already in the guided answer), political persecution, famine, drought, etc.
   (c)  Advantages could include: migrants bring new skills, special skills, migrants are often prepared to do jobs that citizens of the country do not want to do, migrants often want to work hard to earn as much money as they can which means jobs get done quickly and well, there are benefits from cultural exchange – learning new things about different countries.

### 66. EU movements

1  (a)  A second point might include: being able to access health services or education services that are as good or better than those in their home country, or any similar pull factors, or any relevant push factors that you know of.
   (b)  Try to make two developed points.
      •  Schools: Figure 1 talks about the impact of large numbers of migrants on Peterborough's school services, which either means more funding from government is needed to pay for specialists to deal with an unexpected number of non-English as a first language speakers, or it means schools struggling to find ways to meet this challenge on their own.
      •  Health services: have to deal with a big increase in patient numbers and also the same language issues as in schools.
      •  Social services: need to be able to deal with issues such as migrants sleeping rough because they have not been able to find accommodation that they can afford and to make sure children are being cared for properly and families supported.
      •  Cultural impacts: places like Peterborough often did not have a high ethnic minority population before the EU was enlarged, so it can be a culture shock to have shops catering for other cultures, church services changed to meet the needs of worshippers from other countries, schools having to alter what they do to meet the needs of new migrants, and jobs being more difficult to get because of the arrival of well qualified, keen and skilled workers from other countries.

## Changing Urban Environments

### 67. Urbanisation goes global

1  (a)  UK, Botswana, China
   (b)  Other reasons could include: faster natural increase in cities (higher birth rate, lower death rate), economic advantages concentrated in the cities, living conditions and opportunities greater in cities than rural areas.

### 68. Inner city issues

1  (a)  Inner city (as per guided answer).
   (b)  One piece of evidence could be the tower block (a 1970s-style development), the run-down appearance of the area, the fact that the two shoppers are using shopping trolleys and walking rather than shopping in a suburban shopping centre and taking their purchases home in the car.
   (c)  Your answer should refer to the range of environmental, social and economic problems faced by many inner city areas and their connection to the lack of investment in inner city areas, with the knock-on effects these can have on social unrest, crime, etc. answers might also refer to the opportunities to redevelop inner city areas through gentrification as richer people look for somewhere closer to their CBD office.

### 69. Housing issues and solutions

1  (a)  Make the Athletes' Village into affordable homes for local people.
   (b)  Aim to develop the first point and then make and develop a second point.
      •  Point 1: the old housing and the tower blocks mean housing is not pleasant to live in.
      •  Point 2: gentrification – when areas of the inner city become fashionable the rising prices can force poorer people out; or the high demand for housing in the inner city – more and more people cannot afford to live anywhere else and still afford to get to work, but if no new houses are built in the inner city then there is nowhere for these people to go, unless more and more people are crammed into smaller and smaller flats.

### 70. Inner city challenges

1  The question asks you to explain rather than describe, so make sure you say why each point you make would be an advantage or a disadvantage and to whom.
      •  Possible *advantages* of brownfield sites include: helps to revive old industrial areas; services such as electricity, water, gas and sewage are already provided to the site; the site will be close to labour, so if workers are needed during its construction, during the Olympics or after the Olympics they will be available and commuting costs will be reduced; building on brownfield sites reduces the loss of the natural environment that would happen with greenfield site development.

- Possible *disadvantages* include: the expense of clearing the site of old buildings; the very high expense of clearing the ground of industrial pollutants; being surrounded by poorer neighbourhoods may be off-putting; it was in an old industrial area so the transport links required updating.

## 71. Squatter settlements
1  **(a)** Social problems include: drugs, crime, violence and stress. Problems come from so many people living crammed together in difficult circumstances.
   **(b)** Two reasons could include: life in the squatter settlements is still better than life in the countryside; the city offers better prospects; the jobs available in the city pay better; close to friends and family.

## 72. Squatter settlement redevelopment
1  The answer you give will depend on the options you chose.
   - Option 1, *advantages*: the land is cleared and all problems of the site removed as residents are relocated in flats. The land can be sold to developers and more housing or services created which will help the city grow. *Disadvantages*: residents may find they are relocated a long way away from their jobs. The new flats may leave them many storeys up – not a good location if their old house was also their workshop or shop.
   - Option 2, *advantages*: the local community is fully involved, improvements are made that are really needed, costs are kept low and skills are developed within the community. *Disadvantages*: There may be problems with corruption, the work may not be done to a good standard, some residents may get to improve their properties and others not, leading to tensions, and the problems the settlement faces may be too big to be patched over with cosmetic improvements.
   - Option 3, *advantages*: this approach solves the major problems of squatter settlements – unplanned, unserviced and built with poor quality materials. *Disadvantages*: it does depend where the new site is (it may be a long way from jobs, in an unhealthy location) and how big it is. Would such a site simply encourage even more migrants to the city and so swamp the scheme as soon as it was announced?

## 73. Rapid urbanisation and the environment
1  **(a)** growing, water, bypass, repairing.
   **(b)** Effects include: spreads disease; reduces the amount of clean water available to a city; makes an area smell bad and be unpleasant to live in.
   **(c)** Effects include: affecting human health with conditions such as asthma and bronchitis; it creates smogs which reduce visibility and make breathing difficult; it can cause acid rain which damages vegetation and affects the ecosystem when it falls into lakes and ponds; some air pollutants damage the atmosphere.

## 74. Sustainable cities
1  **(a)** 'Recyclable' means that a product can be reused: processed into new products.
   **(b)** Landfill is not sustainable because it cannot be a long-term solution to the waste problem. There aren't enough sites available for much more landfill, so this is not a solution that future generations will be able to use. It does not help reduce carbon emissions because it does not encourage reuse or recycling of products. Landfill can also damage the environment if toxic chemicals escape the site.

# Changing Rural Environments

## 75. The rural-urban fringe
1  **(a)** True, false, true
   **(b)** Examples include: development of leisure activities, new business parks, new transport routes, new housing developments; counter-urbanisation and commuters moving to rural fringe villages; some industries are being moved out of towns and cities because residents no longer want to live near them, such as sewage plants.

## 76. Rural depopulation and decline
1  **(a)** The data shows that wages are lower in Herefordshire than in urban areas in England. This would help explain why people choose to move out of Herefordshire so they can get better paid jobs.
   **(b)** Make one developed point per factor or two points without development. For example: 'An increasing number of people depend on the Internet for entertainment and remote working. As this requires a broadband connection or mobile phone network, rural areas without these services would not be attractive to these people.'

## 77. Supporting rural areas
1  **(a)** To support the rural economy.
   **(b)** Further points may include:
   - The importance of the Internet in allowing some jobs to be done almost anywhere in the country (such as the digital and creative industries mentioned in the extract).
   - The extract talks about special funding for enterprises run by women; this is because the range of jobs available in traditional farming is often felt to be even smaller for women than for men.
   - A lot of the enterprises mentioned in the extract are linked to farming, such as food and drink, agribusiness and forestry, so

the idea might be to make productive links between existing farming and new businesses.
   - Adventure and sports activities show ways in which activities in rural areas are changing towards leisure and tourism, too.

## 78. Commercial farming 1
1  **(a)** The second developed point might include: the ownership of agribusinesses tending to be food processing companies rather than families; the tendency of agribusinesses towards monoculture – the production of just one food type, compared to family farms which tend to have more mixed production; while all non-organic farms in the UK use fertilisers and pesticides, agribusinesses use these in the most scientifically controlled way in order to maximise returns; whereas smaller family farms will have a big role in conserving natural habitats such as hedgerows and patches of woodland, agribusinesses do not often engage in conservation unless their customers demand it or they receive subsidies for it.
   **(b)** This answer will depend on the area of your case study. You should aim to give relevant details and examples in your answer: for example, the name of a large food processing company in your case study area or the effect that the cancellation of a large order had on suppliers of a particular company, or similar. Make linkages between the demands of the supermarkets and food processing companies and what happens in farming as a result.

## 79. Commercial farming 2
1  **(a)** Genetically modified ingredients come from food products that have been genetically engineered to have certain qualities such as disease resistance.
   **(b)** The second point will probably be about organic food being better for the environment and animal welfare.

## 80. Changing rural areas: tropical 1
1  **(a)** As a cash crop.
   **(b)** A second developed point could be: about the soil erosion that cash cropping can cause when it clears the land of vegetation – soil is washed away into streams and rivers and that might mean local people cannot use them for fishing; pesticides from cash cropping may impact on traditional farmers' crops; cash crops could use up a lot of water, leaving less for traditional farmers; there may be conflicts between local farmers and the cash crop agribusiness that is pushing them off their land; there can also be a positive impact: more jobs for local people.

## 81. Changing rural areas: tropical 2
1  **(a)** dam, twelve, increased, 73.
   **(b)** A second impact could be that people who have moved to the cities might send money back to their families, which could help the village do better.

# The Development Gap

## 82. Measuring development 1
1  **(a)** The relationship is a negative (inverse) one, so as income per head goes up, infant mortality rates come down.
   **(b)** Examples include: averages can conceal extreme highs and lows, so a figure like GNI per head can suggest a level of development that is not really true of the country as a whole; development is a complex subject and one measure only records one aspect, for example income per head does not say anything about education levels or about the cultural richness of a country; there are difficulties in recording data from less developed countries, so a mix of measures may give a more accurate picture than just one; some measures are not very good because of changes in global development, for example death rates are now low in almost all countries.

## 83. Measuring development 2
1  **(a)** Ethiopia = low human development.
       Gabon = medium human development.
       Chile = high human development.
   **(b)** The Human Development index is superior because it shows that within the 'poor' South there are countries at different stages of development from low to high, so this classification takes into consideration individual countries.

## 84. What causes inequalities?
1  The two issues are access to enough water and access to water that is safe to drink: quantity and quality. Make connections between water and development, and make links between the points you make. Do not just take straight from the advert.
   - The charity advert says that it takes two hours for Kwame to get water from the waterhole. This means he has no time in the dry season to go to school. If Kwame does not complete his education he has less chance of getting a better job when he is older. This will affect his standard of living considerably. The same is true for many women and children in Ghana who spend most of their dry-season days fetching water. There is no time for the women to attend agricultural school to learn about making their crops do better, so they cannot earn extra from farming, which again affects standards of living right across the country.
   - When Kwame is sick from drinking unsafe water he cannot help on the family farm. If he gets really ill his parents have to pay for him

to see a doctor and pay for the medicine. When a family has so little money, these sorts of expenses can affect their income for the whole year. If Kwame's parents get ill then they cannot work and this can also be very serious for standards of living.

## 85. The impact of a natural hazard

1 **(a)** Two from: fewer people to rebuild Haiti's economy healthcare infrastructure is unable to deal with numbers needing help; most development funding gets spent on rebuilding homes for people.

**(b)** Two from: roads are needed to transport goods and exports; clearing the rubble uses up a lot of development funding; until rubble is cleared businesses and homes cannot be rebuilt; loss of government administration infrastructure makes it harder to focus development where it is needed; loss of schools means children won't get an education.

## 86. Is trade fair?

1 **(a)** By £190 million.

**(b)** Your answer should include one or two of these: schemes that offer a minimum price to cover costs even when world prices fall; schemes that include a premium on the price which is then invested back into community projects; schemes that bring producers together so they have more strength to resist price-cutting measures.

## 87. Aid and development

1 **(a)** Canada = provider; Greece = receiver; India = both; Iraq = receiver.

**(b)** Additional points to add to the partial answer: the majority of countries receiving aid are in South and Central America, eastern Europe and selected Middle Eastern and Asian countries, with the most aid going to sub-Saharan Africa.

## 88. Development and the EU

1 Make sure the information you give uses good details and examples, and make sure they are relevant. Try to identify the interrelationships between factors to show how each of the policies aims to reduce the development gap; for example: 'The CAP gives help to farmers while aiming to prevent environmental degradation so that the very small farms on the EU's periphery can compete with the very big farms in the EU's core.'

## Globalisation

### 89. Going global

1 **(a)** The words in the right order are: cheaper, English, invested.

**(b)** Development of voice services over the Internet. Or any answer that relates to ICT developments making international phone calls cheap, clear and reliable.

### 90. TNCs

1 Your answer should cover both advantages and disadvantages and should make links between factors, and use relevant evidence and examples. Make sure your advantages and disadvantages are for the country itself rather than the TNC, because the question relates only to the country where the TNC has its branches / factories. You can refer to the people as well as the country, though. Consideration can be given to the impacts on the infrastructure as well as the employment opportunities.

### 91. Manufacturing changes

1 Two from:

- Industrialisation in NICs: as more industries develop or move to NICs to take advantage of cheaper labour this will continue the global shift towards manufacturing in countries like China, Brazil, South Korea and India. If other countries became cheaper and more attractive to TNCs, however, manufacturing facilities might shift again to new, even cheaper locations.

- Deindustrialisation in HICs: as manufacturing companies in HICs are unable to compete with the low prices for products from TNCs with factories in the NICs, they either have to move their factories to NICs too or go out of business. This intensifies the deindustrialisation process where manufacturing industries in HICs are replaced by service-sector jobs. However, if deindustrialisation is accompanied by economic recession then some deindustrialised locations might eventually become poor enough to be attractive locations for TNCs to put their factories again (this is known as 're-sharing' – as opposed to 'off-sharing').

- Technological advances in manufacturing: these could be advantageous to HICs because they will often cut down on labour costs which are high in these areas. The resultant increased productivity could counter the advantages NICs enjoy with their lower production costs.

### 92. China

1 Your answer should include the advantages of cheaper wages, restrictions on strikes, reduced health and safety regulations, the importance of government incentives and the creation of Special Economic Zones (SEZs), the importance of China's massive population and the country's relative lack of development prior to the 1970s. Make sure you do not write in vague, generic terms about these factors and that they are always specifically related to China. Try to include relevant details and examples from your case study: for example, overseas investment being allowed in 14 coastal cities, special economic zones

being set up such as Shenzhen, the influence of the one-child policy introduced in 1979 in changing people's behaviour away from having lots of children to buying consumer goods.

## 93. More energy!

1 False – Asia's energy consumption is between 3–10 per cent.
 False – Sub-Saharan Africa *has* shown an increase of 0–3 per cent.
 True – The highly industrialised countries have generally got the lowest increases (all of North America and western Europe).
 True – They are both part of the area with the biggest increase.

2 Asia

## 94. Sustainable energy use

1 **(a)** The USA.

**(b)** Answers might include some of the following points:

- The graph shows that China has seen a very large rise in emissions this century, from around 3 billion tonnes of $CO_2$ in 2000 to around 8 billion tonnes of $CO_2$ in 2010.
- India also has seen emissions rise, but less dramatically, from around 1.2 billion tonnes in 2000 to just under 2 billion tonnes of $CO_2$ in 2010.
- China overtook the US's emissions in 2006, and the US also saw a drop of approximately 0.5 billion tonnes of $CO_2$ in 2008–09, before emissions started to rise again by 2010.
- The UK's emissions have remained pretty much static over the whole time period shown in the graph.

**(c)** 2007

## 95. Food: we all want more

1 **(a)** An increasing number of people need more food.

**(b)** Increase in number of food miles. Use of marginal land leading to environmental degradation. Possible hostilities over demand for irrigation water. Increased rural debt because of the costs of fertiliser and pesticides.

**(c)** The diagram shows that if land is overcultivated or overgrazed, beyond its ability to recover, then it will become less fertile and also soil structure will be weakened so that the topsoil is more vulnerable to being washed away by rain or blown away by the wind. This then makes the land useless for farming. Because marginal land is land that is already risky to farm (due to low rainfall / thin soils / high salt levels), the chance of environmental degradation is already high. Once marginal land is degraded by overcultivation or overgrazing it is also useless for anything else – for example, before farming it might have been used for part of the year as pasture for goats or as a place to gather wild plants and fruits. So the implications are that using marginal land carries a high risk of failure and if people are depending on that land for food, the consequences can be very serious.

## Tourism

### 96. The tourism explosion

1 **(a)** 475 million.

**(b)** 148 million.

2 **(a)** Greater wealth: see guided answer.

**(b)** More leisure time: with paid holidays, people can afford the time to take a foreign holiday.

**(c)** Cheaper travel – this has been very important in opening up international travel to tourists because (a) tourists can afford to fly to their holiday locations; (b) the locations that companies choose to offer cheap flights to can develop their tourism industry rapidly.

### 97. Tourism in the UK

1 **(a)** Germany.

**(b)** Spain.

2 Another common reason is money: people think it is cheaper to have a holiday in their own country as travel costs are lower. You might also have considered: their own country having a well developed tourism industry (as in Spain); or factors like an unfavourable exchange rate which would make a trip to another country more expensive than usual; or the influence of a terrorist attack or air disaster which makes people anxious about flying. Countries also run advertising campaigns to encourage residents to holiday in their own country.

### 98. UK tourism: coastal resort

1 **(a)** The Tourist area life cycle model. It is also called the Tourist resort life cycle model or the Butler model.

**(b)** **U** = Exploration, **V** = Involvement, **W** = Development, **X** = Consolidation, **Y** = Stagnation, **Z** = Decline.

### 99. UK tourism: National Park

1 Make sure that you have definitely used a UK National Park example or a UK coastal resort example, and that you write the name of your example clearly. Link your points together and use relevant details with relevant examples. For the Lake District, you could make this link: 'The Lake District has many beautiful mountain walks, such as the climb up the Langdales, and this attracts many visitors who enjoy outdoor activities such as climbing and walking.' Try to include some geographical terms, for example: 'The Lake District has many *honeypot locations* such as Tarn Hows or Dove Cottage, which attract the most visitors of all because they are especially beautiful or historic.'

2 The details of the plans here will depend on your chosen area, but

make sure you describe two as the question asks. The plans should be linked to specific problems, for example: 'The Lake District National Park Authority (LDNPA) is concerned that younger people want more active and challenging activities than hiking and walking, which could mean a drop in visits by younger people. So the LDNPA is developing mountain bike trails in forested areas to attract the next generation of visitors to the National Park.'

## 100. Mass tourism: good or bad?
1   Link your reasons to evidence from the photo or from your own knowledge. Try to add some relevant detail or a brief example if you know one. For example, you might want to talk about the environmental impact of erosion caused by tourists. Link this to the evidence in the photo that tourists in Kenya are driven around the savannah grasslands in minibuses, which cause a lot of erosion, particularly when it is wet and they churn up the ground into mud which then dries and blows away as dust. You could add a point that is not directly related to something in the photo, for example the damage done to the environment by tourists flying from their home country to tourist destinations and home again. Most tourists come to Kenya from the UK, the US, Italy and Germany, all of them a long way from Kenya, and the number of tourists coming from India and China is also increasing, so this impact may only get worse in the future.

## 101. Keeping tourism successful
1   (a)   Beach holidays and wildlife safari holidays.
    (b)   Cultural tourism is when tourists come to experience the history and traditions of a country or region.
    (c)   Your answer might go on to say: so the Kenyan government would want to expand the number of places in Kenya tourists travelled to so that it could reduce the negative impacts on existing tourist areas while still increasing the total number of tourists who come on holiday to Kenya. If negative impacts such as erosion, overdevelopment, pollution and conflict with local people over resources were allowed to get worse, tourist numbers could drop quickly, which would be very bad for the Kenyan economy. Spreading tourism out over more areas in Kenya will also mean more of the economic benefits of tourism reach more of the Kenyan people. This could happen through more Kenyans getting jobs in tourism, but also if cultural tourism was developed in different areas then more Kenyans could be employed making handicraft products to sell to tourists. This will stop all the investment in infrastructure improvements being concentrated in the holiday areas with the rest of the country not benefiting.

## 102. Extreme tourism
1   (a)   The extract mentions a limit on the numbers of people from any one ship that can land on Antarctica on each trip to the shore. If unlimited numbers of people were allowed to land then it would be very hard to control who went where. There would be a lot more feet to trample vegetation, there would be a much higher risk of wildlife being disturbed and it would be much more difficult to make sure no litter of any kind was left behind. The 100 people guideline is also important more generally, since most tourists will want to travel on smaller ships so they have a better chance of getting to go on shore each trip. This means that tour operators are more likely to run trips on smaller ships, which can take fewer people, so overall the numbers of tourists who can go to Antarctica during its short spring and summer season is reduced.
    (b)   Extreme environments are very fragile, so they are very easy for tourists to damage and the damaged area takes a very long time to recover. They are fragile because life is really difficult in these environments and the plants and animals survive on a knife-edge: any change and their specialist adaptations don't cope well enough to survive. So if tourists trample plant life, drop litter or disturb wildlife, the impact on the ecosystem can be really serious.
    (c)   This answer will depend on the case study you have done. For the Antarctic you could name the Treaty of Antarctica, the Polar Code (not yet completed), the MEPC's 2011 treaty extension banning the use of heavy fuel oil in the Antarctic area.

## 103. Ecotourism
1   (a)   False, true, false.
    (b)   The difference between conservation and stewardship is not massive but it is very important: conservation is about protecting and managing the environment while stewardship is about a responsibility for caring for the environment, a responsibility to conserve it, to commit yourself to putting its needs above everything else.
    (c)   The tourists that would be attracted to an ecotourism holiday are likely to be people who are concerned about the impact of mass tourism on the environment. They are likely to have an interest in

visiting unspoiled and exciting ecosystems, such as rainforests, deserts, etc., without damaging them. They may want to have a closer link with local people in their holiday destination and might be interested in learning new things from people living in a different way from them. They might want their holiday to benefit local people rather than big international companies.

## Exam Skills
### 104. Stimulus materials – an introduction
1   (a)   Make sure you complete graphs like this as accurately as you can – use a ruler to measure where to put the 110 line and draw the sides and top of your bar really neatly and with straight edges.
    (b)   Be careful not to just lift figures from the stimulus material without showing how they help answer the question. For example, saying 'The number of shops declined from 254 to 110 …' is OK but if you added '… which makes it the type of service showing the greatest loss', then you are showing that you appreciate the overall pattern of the data. Saying 'There are 14 fewer Post Offices in 2011 compared to 1981' is better than repeating the data: 'The number of Post Offices has gone from 36 to 22', because it shows that you can manipulate the figures. Saying 'The number of petrol stations declined between 1981 and 2001 but has remained unchanged since then' shows that you recognise a trend and how services have changed over time.

### 105. Using and interpreting photos
1   *Reason for visiting*: wildlife (these are Adelie penguins), landscape, icebergs.
    *How tourism could damage Antarctic ecosystems*: scaring penguins during breeding season, pollution from boat engine (spilled diesel in the sea, for example), dropped litter, damage to fragile plant life.
    *How tourism is controlled*: small numbers only can visit at any one time (two boats are shown here), controls over where people can go (the group on the beach are not being allowed to wander but are in a tight group, apart from the photographer).

### 106. Labelling and annotating
1   Aim to include four labels. Evidence from the photo could include: the presence of a small meltwater lake at the snout of the glacier; the evidence of lateral moraine in the foreground of the picture, which suggests the glacier was recently wider and further forward; the bare ground (no vegetation) around the snout of the glacier, suggesting that this area has only been recently exposed (there is sparse vegetation on the lateral moraines); the absence of a terminal moraine, suggesting retreat rather than a glacier that has been pushing forward. Your labels only need to identify these features rather than explain them.

### 107. Graph and diagram skills
1   (a)   This should be halfway between the 140 and 160 lines on the vertical axis and directly above the 2011 point on the horizontal axis.
    (b)   70 per cent.
    (c)   Eight years (between 2000 and 2008).

### 108. Map types
1   (a)   Brazil should be shaded as an upper-middle income country, Kenya as a low-income country, Saudi Arabia as a high-income country and Ukraine as a lower-middle income country.
    (b)   Choropleth.

### 109. Describing maps
1   Most hot deserts occur close to the tropics of Cancer and Capricorn. The greatest extent of hot desert is in a belt across North Africa and the Middle East. There are no hot deserts in the higher latitudes of the northern hemisphere, although the map shows hot deserts stretching far to the south in South America. Hot deserts in the Americas are most frequently located along the eastern coastline.

### 110. Comparing maps
    (a)   About 135 m.
    (b)   300 m

### 111. Exam skills
1   •   Describe – Write about what you can see.
    •   Explain – Write about the reasons for something; explain why.
    •   List – Make a list.
    •   Compare – Write about the similarities and differences.
    •   Contrast – Write about the differences.
    •   Outline – Write down the most important points.
    •   Annotate – Add labels with details.
    •   To what extent…? – Come to a conclusion after giving different points of view.
    •   Use a named example… – Use a case study in your answer.

Published by Pearson Education Limited, Edinburgh Gate, Harlow, Essex, CM20 2JE.

www.pearsonschoolsandfecolleges.co.uk

Text and original illustrations © Pearson Education Limited 2013
Edited, produced and typeset by Wearset Ltd, Boldon, Tyne and Wear
Illustrations by Wearset Ltd, Boldon, Tyne and Wear
Cover illustration by Miriam Sturdee

The right of Rob Bircher to be identified as author of this work has been asserted by him in accordance with the Copyright, Designs and Patents Act 1988.

First published 2013

17 16 15 14 13
10 9 8 7 6 5 4 3 2 1

**British Library Cataloguing in Publication Data**
A catalogue record for this book is available from the British Library

ISBN 978 1 447 94089 0

Printed in Slovakia by Neografia

**Acknowledgements**
The publisher would like to thank the following for their kind permission to reproduce their photographs:
**Alamy Images**: 1Apix 44, Ashley Cooper Pics 20, BrazilPhotos.com 80, Cath Harries 69, Chris Howes / Wild Places Photography 36, CuboImages srl 6, David Bagnall 37, FirePhoto 8, geogphoto 14, Ian Canham 68, imagebroker 11, Janine Wiedel Photolibrary 65, JLImages 4, Joe Bird 70, Olivier Asselin 84, Paul Glendell 12, Paul White Aerial Views 75, Pictorial Press Ltd 47, Stephen Dorey – Gloucestershire 74, Stephen Foster 56, Tommy Trenchard 85; **Getty Images**: Michel Setboun 105; Science Photo Library Ltd: Dr. Juerg Alean 106, Gary Hincks 13, University of Dundee 19; **Shutterstock.com**: Natursports 100; **USGS**: Carl Key 43cl, Dan Fagre 43cr, Lindsey Bengston 43r, T. J. Hileman, Courtesy of GNP Archives 43tl
(Key: b-bottom; c-centre; l-left; r-right; t-top)

**Figures**
Weather chart on page 18 from http://www.metoffice.gov.uk/media/pdf/a/t/No._11_-_Weather_Charts.pdf

**Maps**
Maps of Castleton, Langdale and Hurst Castle Spit. Reproduced by permission of Ordnance Survey on behalf of HMSO, © Crown Copyright 2013. All rights reserved. Ordnance Survey Licence number 100030901 and supplied by courtesy of Maps International.

**Text**
Page 30 extract from http://www.guardian.co.uk/environment/2012/sep/03/ecuador-yasuni-conservation, Guardian News and Media Ltd; Extract 2. adapted from http://archive.defra.gov.uk/rural/documents/economy/regr-rural-growth-networks.pdf

All other images © Pearson Education Limited

Every effort has been made to contact copyright holders of material reproduced in this book. Any omissions will be rectified in subsequent printings if notice is given to the publishers.

In the writing of this book, no AQA examiners authored sections relevant to examination papers for which they have responsibility.